Animals of Moorland and Forest

World of Wildlife: THE NORTH

Animals of Moorland and Forest

ORBIS PUBLISHING · LONDON

From the original text by Dr Félix Rodríguez de la Fuente
Scientific staff: P. de Andres, J. Castroviejo, M. Delibes, C. Morillo, C. G. Vallecillo
English language version by John Gilbert
Consultant editor: Dr Maurice Burton

Printed in Italy by IGDA, Novara
ISBN 0 85613 494 5

Contents

Acknowledgments

J. Amos/Photo Researchers: 122, 126
R. Austin/F. W. Lane: 209, 217
Barnaby's Picture Library: 282
H. M. Barnfather/Bruce Coleman: 221, 245
Des Bartlett/Bruce Coleman: 201
F. Bel-G. Vienne/Jacana: 30, 31, 35, 165, 175, 207, 270, 271, 283, 292
A. Bertrand/Jacana: 147, 151, 172
F. Blackburn/Bruce Coleman: 218, 265
R. Blewitt/Ardea Photographics: 259
Bos/Jacana: 256
H. Bristol jr./Photo Researchers: 142
M. Brosselin/Jacana: 40, 104, 108, 208
J. Burton/Bruce Coleman: 40, 49, 56, 134, 172, 191
M. Celo/Jacana: 298
Chevalier/Jacana: 195
D. Choussy/Jacana: 31, 32
A. Christiansen/F. W. Lane: 162, 228, 267, 274
Bruce Coleman: 6, 118
J. Costa/Edistudio: 3, 140
L. M. Cox/Photo Researchers: 187
W. Curtz/Ardea Photographics: 224, 239
C. R. Dawson/Bruce Coleman: 87, 91
A. J. Deane/Bruce Coleman: 38, 275
Edistudio: 60, 62, 65
J. Elosegui: 27, 39
A. Fatras/Jacana: 51, 58, 78, 117, 190, 191, 192, 198, 230, 231, 239, 261, 263
R. Fenaux/Jacana: 240
P. Fornaison/Jacana: 89
Frederic/Jacana: 91
Y. Gaugis/Jacana: 109
S. Gillsäter/Tiofoto: 171, 197, 286, 287, 290, 291, 295
S. Gooders/Ardea Photographics: 16
A. Gutiérrez/Edistudio: 27
B. Hawques: 9, 26, 53, 55, 169, 223
P. Hinchcliffe/Bruce Coleman: 261
G. Hollis/Photo Researchers: 125
E. Hosking: 37, 74, 86, 91, 99, 246, 256, 259, 273
Jeager-Kay/Photo Researchers: 132
R. Kinne/Bruce Coleman: 80, 180, 199, 278
R. Kinne/Photo Researchers: 137, 139, 199, 283, 287
S. Krasemann/Photo Researchers: 185
J. Lanceau/Jacana: 55
J. Lindblad/Bruce Coleman: 218
Longo: 78
B. Mallet/Jacana: 143
A. Margiocco: 45, 56, 209, 217
J. Markham: 47, 257
D. Middleton/Bruce Coleman: 227, 268
A. Moliner/Jacana: 23, 227
P. Montoya/Jacana: 117, 147, 271, 293
C. Mylune/Ardea Photographics: 107
Nadeau/Jacana: 169
G. Oddner/Tiofoto: 131
Okapia: 180, 187, 213, 226, 239, 280
Ch. J. Ott/Bruce Coleman: 58, 214
X. Palaus: 75, 82, 130, 132, 133, 139, 141
E. Park/Bruce Coleman: 235
L. Pechuan: 23, 39, 253
R. Peterson/Bruce Coleman: 274
H. Phetschringer/Bavaria: 80, 149, 152, 186, 228
J. Pons: 18, 103
S. C. Porter/Bruce Coleman: 77, 86, 261
H. Reinhard/Bavaria: 80, 149, 152, 186, 228
H. Reinhard/Okapia: 151, 157
H. Reinhard/Zentrale Farbbild Agentur Gmbh: 144, 160, 182, 241, 277
S. Roberts/Jacana: 9
F. Rodríguez de la Fuente: 73, 94, 96, 99, 105, 115, 236, 275
F. Roux/Jacana: 177
A. Sarró: 13, 14, 26
W. Schramil/Jacana: 277
Schrempp/Okapia: 155, 177
H. Schüneman/Bavaria: 190, 250, 251
A. Silva/Safoto: 111
H. W. Silvester/Bavaria: 194
J. Six: 66, 93, 289
R. T. Smith/Ardea Photographics: 204
P. Summ/Jacana: 165
X. Sundance/Jacana: 297
R. Tercafs/Jacana: 285
J. F. Terrasse: 211
J. F. Terrasse/Jacana: 105
W. Tilgener/Jacana: 248
Vala/Jacana: 3, 13
J. P. Varin/Jacana: 75, 103, 105, 156, 158, 161, 166, 296
J. Viellard/Jacana: 11
J. Vilanova/Edistudio: 124, 130
A. Visage/Jacana: 84, 214, 236, 243, 255
J. Wightman/Ardea Photographics: 68, 73, 87
J. V. Wormer/Photo Researchers: 197
Zentrale Farbbild Agentur Gmbh: 37, 42, 159, 235, 246
D. Zinguel/F. W. Lane: 256

Foreword

The trees of the deciduous forests of temperate Europe, Asia and North America, although less massive than those of the tropical rain forest, are incomparably more beautiful. The delicate green of the spring foliage is transformed in autumn into a magnificent tracery of brown, yellow, orange and red.
For the animals of wood and forest, life is conditioned by the contrasts of the seasons. Spring marks the return of migrating birds from warmer lands in the south, the reappearance of burrowing and hibernating mammals. By summer new generations share the joys and hazards of forest life. Autumn, with its harvest of fruit and nuts, is the season of plenty, but already preparations are afoot for the cold, difficult months ahead. Then, after the bleak desolation of winter, life returns in all its cheerful splendour.
Less spectacular than the great carnivores of the jungle, the predators of forest and moorland – bear, lynx, fox, weasel, polecat and marten – rely more on cunning and agility than on brute strength. Safe from their marauding are the lordly red and roe deer with their branching antlers and the sturdy wild boar with their dangerous tusks. But there are innumerable rodents and nesting birds at their mercy.
The birds of the temperate woods and grasslands are fascinatingly varied. Eagles, kites, falcons, kestrels, buzzards, goshawks, sparrowhawks and owls, all with individual hunting techniques, pose a constant threat from the air. The game birds – bustards, partridges and quails – perform colourful and noisy courtship rituals. And the glades and treetops echo to the joyful song of tits, warblers and nightingales.
The secrets of the moorland and forest animals of the northern hemisphere are dramatically revealed in word and picture in this, the fifth volume of *World of Wildlife*.

World of Wildlife

The North: Animals of Moorland and Forest

CHAPTER 1

The Spanish lynx: hunting cat of the south

In the mountains of the Iberian peninsula the vegetation is wild, tangled and colourful. Breaking the almost impenetrable carpet of heather and cistus, clumps of arbutus, lentisk and broom intermingle; and the banks of rushing streams are lined with oleander and bramble. Here, in regions where man has seldom ventured, are tall, venerable oaks and cork oaks; and in this magnificent, savage landscape, many animals roam. Of the hunters none is more imposing than the Spanish lynx (*Lynx pardellus*), largest feline of the Mediterranean fauna.

This splendid carnivore, like others of its relatives, is a wonderfully versatile animal. It can leap with a single springing bound from a narrow rock ledge to the higher branches of a tree, or it can glide silently and stealthily through the underbrush in pursuit of any form of prey, ranging from a tiny mouse to a large deer. The short hair of its coat is ideal for slipping effortlessly through the thorny undergrowth and the black spots against the yellowish-brown background help to conceal outlines in the interplay of woodland light and shade. The short, stumpy tail is not as mobile or well developed as that of other felines and thus does not betray the hunter's presence or provide a key to its passing moods.

Powerful muscles control the legs, and the soft padding of the feet furnishes a perfect springboard for jumping, likewise deadening the telltale sound of snapping twigs and loose pebbles. The claws are long and sharp, retracted in their sheaths when not in use but as lethal as sword points when employed against a victim that may be larger than the predator itself. The jaws, with their deeply rooted teeth, are equally formidable weapons, often completing the deadly work begun by the claws. Nor are the distinctive cheek ruffs mere adornment, for they have the effect

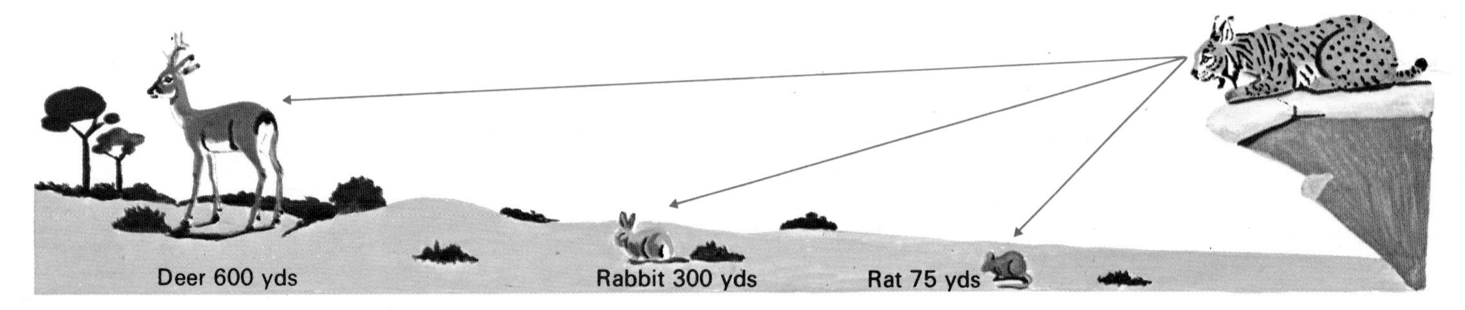

The naturalist Waldemar Lindeman conducted an interesting experiment to verify the Spanish lynx's reputedly keen vision. Placing the predator on a high ledge overlooking the plain, he arranged for different types of prey – a rat, a rabbit and a deer – to be brought into view at varying distances. The lynx had no difficulty in spotting them the moment they moved, at 75 yards, 300 yards and 600 yards respectively.

of enlarging or contracting the contours of the face and of blurring its outline as the lynx lies in ambush for prey.

The Spanish lynx has the added natural advantage of being a medium-sized animal so that it is not at any time too conspicuous from the viewpoint of the prey it is tracking through the dense vegetation of swamp and mountainside. The male measures 32–44 inches from muzzle to tip of tail and its weight ranges from 25 to 45 lb. Some individuals have been known to top the latter weight but this is rare. The female is smaller and seldom weighs more than 30 lb.

'Lynx-eyed' is a proverbial phrase and the animal was in fact named after the Argonaut pilot Lynceus, renowned for his abnormally sharp vision.

The German naturalist Waldemar Lindeman, who has specialised for many years in studying the physiology and behaviour of this lynx, tested the species' hunting prowess and in particular the keenness of its eyesight by rearing two newly born animals and training them to pursue the same small mammals – mice, rabbits and hares – as they would have been taught by their mother to do in the wild. He placed one of the young lynxes on a rocky headland commanding an uninterrupted view of a wide area of level ground. On receipt of a light signal, Lindeman's assistant pulled a cord to which had previously been attached various types of familiar prey. As soon as the feline caught sight of one of these animals it became noticeably restless, flexing its muscles in readiness to pounce. Further experiments showed that the lynx could see a rat at a distance of 75 yards, a rabbit at 300 yards and a roe deer at 600 yards. This is a remarkable achievement since few mammals are capable of distinguishing an object, even a moving one, over such distances.

The lynx and its territory

The Spanish lynx lives in regions where there is plenty of thick vegetation, which explains why it is not often seen in the wild. Its cryptic coloration and instinctively secretive nature are additional reasons why naturalists find it difficult to study the animal in its natural surroundings. Observations made in captivity, however, together with clues such as scent posts and signs of fighting, indicate that the animal is a decided individualist which leads a solitary existence at all times apart from the breeding season and normally roams a territory of about 3–4 miles. Throughout this area the prowling feline takes care to cover its excrement with sand or earth, but does not take the same precautions when it urinates. In fact it makes use of its

urine to stake out territorial boundaries, spraying it accurately over selected landmarks, without taking the trouble to adopt a special posture. As it ranges over its terrain it systematically sprays bushes, tree trunks and rocks, serving clear warning to potential rivals and enemies that these vantage points are already occupied.

The lynx can also signal its presence by visual means, ripping the bark off certain trees which are apparently used as places of concealment and shelter.

It may happen that in spite of all these precautions two lynxes will clash on disputed territory, in which case the confrontation may be quite violent, the drama heightened as both rivals howl in unison–the volume of sound being unusually powerful for animals of this modest size. The long cheek ruffs play an important defensive role in these intraspecific contests. Flailing claws and bared fangs that might otherwise inflict fatal wounds on face and neck bury themselves harmlessly in the thick ruffs of hair growing from either cheek.

Outside the breeding season the lynx will seldom tolerate the company of another member of the species on its territory, let alone a rival predator. When the lynxes were far more numerous than they are today, they would fiercely attack foxes, wildcats, badgers, mongooses and otters, all of which shared the same habitat and quickly learned to give the quarrelsome, aggressive carnivores a wide berth. Now that these wild species are far less abundant, lynxes are sometimes compelled to go hunting around farmyards, where they often clash with guard dogs.

It is obvious, therefore, that in its traditional habitat the Spanish lynx has always played a vital role in the food chain, one of its principal functions being to control the numbers of smaller carnivores. Among the latter, the fox seems to have been picked out as a favourite victim, largely because it was the most formidable of rivals for food. Since the fox was never an equal match for the lynx in face-to-face combat, it sensibly decided, whenever possible, to slink off to less dangerous districts rather than provoke a fight which could have only one result. Even in their new habitats the foxes would seldom pose a local threat because there were still enough lynxes to prey on them. But when the lynx population itself began to decline–for all the usual reasons–the foxes multiplied. Soon farmers and hunters were up in arms and eventually, in 1960, the Spanish authorities decided to try to restore the super-predator to its former place by granting it official protection. How far this attempt to conserve the lynx will be successful remains to be seen. It will be years before it once more begins to exert a significant influence on the natural balance in its habitat.

Ambush and assault

The lynx roams every square inch of its territory until it is familiar with every bush, every rock, every tuft of grass. Sometimes it will be seen squatting for hours on end, in broad daylight, on one of the mounds where rabbits habitually deposit their droppings. Although it appears lazy and relaxed, the lynx

Facing page: The beautiful Spanish lynx has long been hunted for the sake of its fur and is today threatened with extinction. Only a few animals now roam wild in most parts of southern Europe but there are still some 150 pairs in Spain, some of them protected in the Doñana reserve where scientists are able to study their daily and seasonal habits.

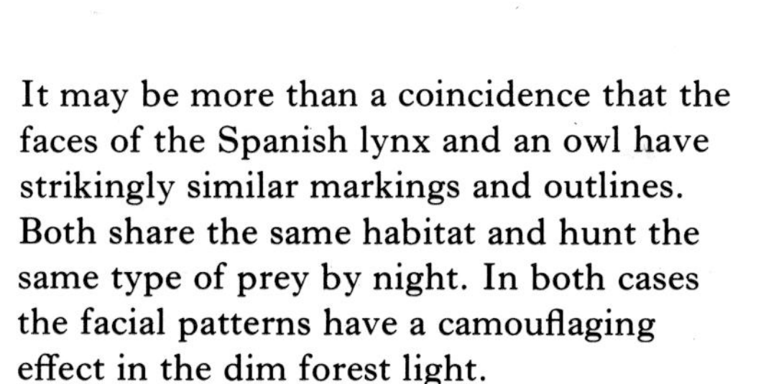

It may be more than a coincidence that the faces of the Spanish lynx and an owl have strikingly similar markings and outlines. Both share the same habitat and hunt the same type of prey by night. In both cases the facial patterns have a camouflaging effect in the dim forest light.

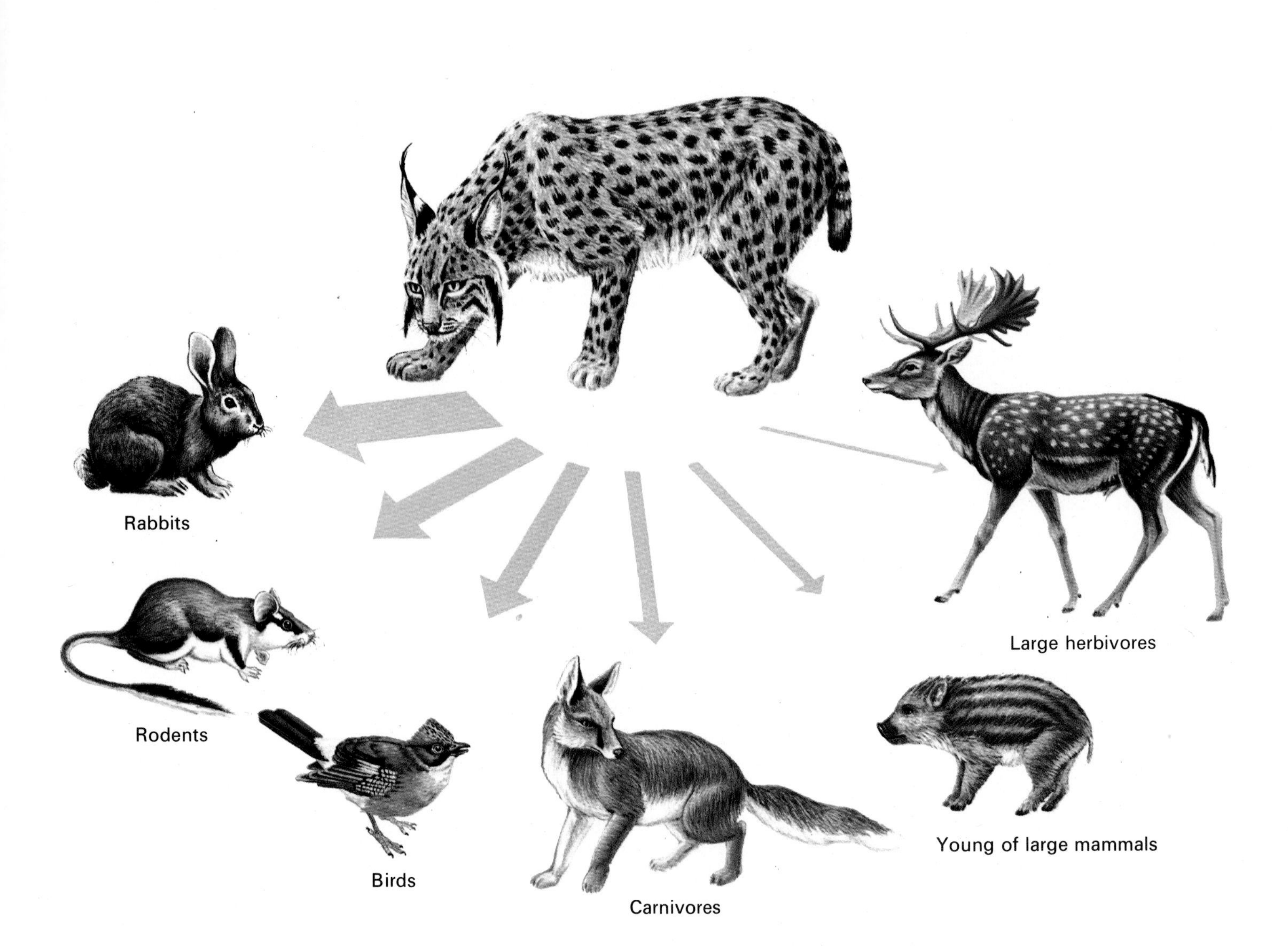

Comparative food preferences of the Spanish lynx.

is really alert and wary, trembling at the slightest sound. At other times it may be found at the foot of a cork oak, casting keen eyes over the branches, listening for give-away cracking noises; or it may wait patiently for muskrats to appear at the edge of a pond or stream.

When a wild boar is about to give birth a lynx may lurk in the vicinity, ready to drop from a tree on an incapacitated female or thread its way through the undergrowth to find babies that have been left temporarily unguarded by the mother.

As a rule, however, the lynx hides itself in dense, tangled vegetation, making its principal lair in a rock cleft or hollow tree trunk. Alternative hiding places occasionally come into use but are not occupied for any length of time. Like most of its feline relatives, the lynx is active both by night and day. In areas where it has enemies to guard against, darkness is obviously a great advantage; but in regions where it has nothing to fear, it moves around freely at any time. To some extent its hunting expeditions are dependent on the season, with daytime excursions much more common in winter than in the heat of summer. There is really no other limiting factor. With its remarkable eyesight, acute hearing and sensitive whiskers, the lynx is equipped to hunt in almost total darkness just as effectively as in bright daylight.

The easiest animals to catch in the darkness are those that make a noise while they are moving about or feeding, notably herbivores and rodents. The lynx listens carefully for telltale sounds and identifies them unerringly. It knows, for example, that sharp crackling noises are made by wild boars as they rummage in the ground beneath an oak, and it steers clear of these too formidable opponents. On the other hand, the sound of a dormouse nibbling at a hazelnut will attract the predator's attention immediately and stir it to action. Silently, yet with surprising speed and agility, the lynx glides like a shadow through the brambles towards the imprudent victim. Once within striking distance, the end is sudden and quick. Now and then the frantic squeal of a trapped rabbit shatters the night air. This is a signal for the cat to interrupt whatever it is doing and to bound off in the direction of the sound, which may betray not only the presence of an easy victim such as a fox or badger, but possibly that of another lynx which is hunting on forbidden ground. In the latter situation the lynx runs the risk of emerging defeated from the encounter and being forced to abandon its territory.

The lynx's territorial instinct has often proved its undoing. In those parts of the maquis where poachers have laid traps for rabbits, a lynx, drawn irresistibly to the spot by an alarm cry, may fall straight into the snare prepared for its victim. This has been one of the reasons for their gradual disappearance in some regions, whereas the more cunning foxes, knowing how to avoid such traps, profit from the upset balance. There is really no justification for trapping now that myxomatosis has so decimated the rabbit population, and it is high time that it was curtailed altogether, particularly in those few regions where the Spanish lynx still runs wild. If things are allowed to continue as they are, the inevitable result will be the extinction of yet another European animal, one of unusual interest and singular beauty.

Geographical distribution of the Spanish lynx.

SPANISH LYNX
(Lynx pardellus)

Class: Mammalia
Order: Carnivora
Family: Felidae
Total length: 32–44 inches (80–110 cm)
Length of tail: $4\frac{3}{4}$–$5\frac{1}{4}$ inches (12–13 cm)
Height to shoulder: 20–28 inches (50–70 cm)
Weight: 25–45 lb (12–20 kg)
Diet: small ungulates, rabbits, rodents, birds
Gestation: 65–72 days
Number of young: 1–4, usually 2
Longevity: 10–15 years

Adults
Yellowish-brown coat, freely scattered with black spots. Short tail, rounded at tip. Thick cheek ruffs and long tufts of hair on ears. Female smaller, seldom weighing more than 30 lb.

Young
Coat lavishly spotted but cheek ruffs and ear-tufts less noticeable.

Lords of the Doñana reserve

In Spain the lynx is thought to be still present in three widely contrasted regions—in the marismas (Guadalquivir swamplands), in the central mountain chains and among the peaks of the Pyrenees. It is not absolutely certain, however that the animal any longer roams the Pyrenees.

The community inhabiting the Guadalquivir delta is more or less restricted to the interior of the Doñana reserve and the surrounding country; but although the spacious park certainly affords protection to this species, there are estimated to be no more than fifteen pairs.

The reserve, once a private hunting precinct, is efficiently patrolled by wardens; and because of its moderate climate and smooth terrain, it is an animal wonderland, visited by scientists the world over. The lynx is one of the prize exhibits and the authoritative work on the species by the reserve's director, Dr Valverde, has provided the basis for the following information.

The Spanish lynx roams the low-lying maquis, where the Montpellier cistus blooms, as well as open pine woods with thick underbrush and—occasionally—swamps. It also frequents oak

The Spanish lynx has a meticulous hunting pattern which involves comparatively little effort. For a start the feline simply sits absolutely still, watching and listening. Having spotted a victim, it creeps stealthily through the grass, belly close to the ground, until it comes within range. Finally the predator pounces, claws unsheathed and jaws agape.

woods where cistus, lentisk, wild olive and myrtle grow, and it is here that it concentrates mainly on catching rabbits. To do this it uses two types of hunting technique: lying in wait, and stalking the prey. Initially it squats on its haunches in a carefully selected spot–usually a field or piece of open land–close to the entrance of a burrow. There it waits patiently, body immobile apart from an occasional inclination of the head, drawing no attention to itself.

When it eventually sights its victim, no matter what the distance separating them, the lynx crouches down, never lifting its eyes from the objective, and begins to move stealthily forward through the grass, belly to the ground. Even though it advances at a surprisingly brisk rate, the animal's spotted coat is hard to distinguish in the dappled light, so that it can approach as close as 10–15 feet from its prey without being noticed. Once within range it pounces, giving the startled rabbit little chance of escape. If the rabbit does manage to evade the predator's clutches, the lynx will not normally pursue it for long. In the majority of cases, fangs sink deeply into the victim's neck.

Other favourite prey are ducks. Numerous observations in the Doñana reserve indicate conclusively that the lynx has little fear of water and is in fact a surprisingly good swimmer. Although partridges are very numerous in the reserve, the lynx seldom elects to hunt them. Now and then, however, it will pluck one of these stout game birds out of the air as neatly as its relative, the caracal, traps a sand-grouse.

Alone or with a companion, a lynx may descend on a herd of deer during the breeding season, selecting any one of the several hundred fawns that are born on the reserve every year, or perhaps an elderly doe or otherwise enfeebled adult. To kill an animal of this size entails a much greater effort on the part of

Facing page : The spotted coat of the Spanish lynx blends perfectly with the background vegetation, allowing it to stalk a victim and catch it by surprise. Swamps and marshes not only abound in prey but also provide effective hiding places, seldom explored by enemies, man included.

Defence of territory, often gives rise to fierce fights between rival lynxes. Yet such combats follow a prescribed ritual and are often interrupted by diversionary actions, including mutual rubbing of foreheads.

the carnivore, for in addition to sinking the teeth into the throat, enough pressure has to be exerted by the jaws to bring about strangulation. In general the large ruminants are left alone, for not only may the struggle be too difficult and dangerous for the feline, but such a victim carries more meat than is needed. In the ordinary way, a couple of pounds of flesh is quite sufficient for the carnivore's daily requirements. Hunting smaller animals is less complicated, less tiring and far safer.

Unlike most members of the feline tribe, the Spanish lynx never consumes any prey that it has not itself killed. Most felines are not averse to carrion or taking the kills of other carnivores.

Another strange habit of this predator is that it never feeds in the place where it has attacked and killed its prey. A small carcase will be carried away in the mouth, a larger one dragged off, feet trailing. A Doñana warden once watched a lynx pulling along a young deer for some 150 yards and another carrying a dead rabbit for more than half a mile, short tail waving and head held high to avoid letting the victim touch the ground. When it reaches the site chosen for the meal the lynx eats as much as it can manage and studiously buries the remains, after digging in the sand and stirring up leaves and grass with its forepaws. When the victim is comparatively large, this involves considerable effort, but the carnivore still goes through the motions, even if it only manages to bury a part of the carcase.

The Spanish lynx often attacks animals considerably larger and heavier than itself, notably deer which are too young, too feeble or too old to defend themselves. In such cases the predator launches a surprise attack, holding the victim firmly with its claws and then burying its fangs in the throat with a view to strangulation. An animal killed in this way often bears four neat punctures at the base of the neck – the marks of the predator's sharp canines.

Life in the lowlands

The easiest time to study the day-to-day life of the Spanish lynx in the Doñana reserve is during January, which is the mating season. Observation is more difficult between mid-February and early May; during this period one may by chance see a gravid female or later, a mother lynx accompanied by her young, but such encounters are rare.

By July there are quite a few babies, youngsters and adults to be seen, and between August and mid-November their appearances are increasingly frequent, affording good opportunities for close study of behaviour. But shortly before the breeding season the animals become noticeably more shy and retiring.

During the summer the lynxes can sometimes be seen shortly after daybreak, moving out on their hunting forays between 8 and 9 o'clock and apparently making their kill within the next hour or so. By then the sun is high and the heat already intense, so that the normal procedure is for the animals to rest in the

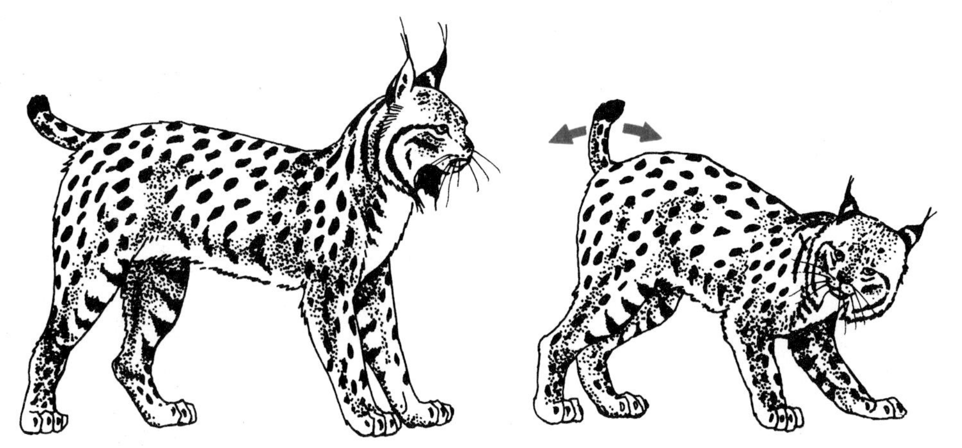

Solitary for most of the year, lynxes come together in the mating season early in the new year. The female usually initiates the courtship ritual and signifies her sexual interest by flexing the forepaws, raising her rump and waving her tail from side to side.

shade until late afternoon. From about 6.30 until dusk they are once again active and this is usually the best time of day to accumulate scientific information.

The daily timetable is quite different in winter. At this season it is late in the morning–around 11 o'clock–that a lynx is most frequently seen, perhaps peacefully stretched out on the ground. In fact the end of the morning and the early afternoon are the hours normally devoted to hunting. Towards 6 o'clock, with night falling fast, the animals once more embark on their quest for food.

Lynxes generally mate early in the new year but this is not an absolute rule, for with a gestation of 65–72 days, some females have already produced kittens by January, indicating that they must have been sexually receptive as early as November. It is, however, more usual to see litters in March and April. The males, having led a solitary existence for the rest of the year, seek their partners when winter comes, emitting characteristic cat-like calls. Each female evidently prowls round the outskirts of her future mate's territory and then, scenting him, ventures into his hunting grounds and settles down with him for the remainder of the mating season. After they have coupled, she retires to her own territory.

Two months later she gives birth to a litter of one to four kittens (normally two) in a tree hollow, in dense undergrowth, in a badger's sett or even in the old nest of a stork. The babies are born blind (like most carnivores) and are quite incapable of doing anything for themselves. Their growth is very slow and only around the ninth or tenth day do they open their eyes. Five weeks after birth they venture out on short exploratory rambles, but it is another month before they accompany their mother on serious hunting expeditions and begin to learn their trade.

The mother stays close to her youngsters during their infancy and adolescence but when the mating season comes round again they have to face their first real challenge. Rejected by the adults, now busy with love-making, they are compelled to find unoccupied territory of their own. But in this second year of life the young lynxes are already capable of indulging in sexual activity and are themselves potential reproductives.

Despite the common assumption about territorial inviolability, Dr Valverde has recently produced evidence to suggest that such boundaries are not all that sacrosanct. He describes how several adult lynxes may be found occupying the same stretch of ground at any time of the year. If this is typical behaviour, it would

Like most cats, lynxes are exceptionally vocal when on heat, shattering the peace of the Doñana reserve with their loud howls as they call to their mates.

The rigid tufts of hair adorning the ears of the lynx are very sensitive. Scientists studying captive animals have noted with interest that these long hairs are often used to flick irritating flies off the back, something that other animals do with their tail. The stubby tail of the lynx, however, is too short to cope with this problem.

Facing page : Alert to the slightest noise, even when it is resting, the Spanish lynx patrols its territory both by day and night. Once stirred to action, the feline moves silently and swiftly on soft, padded feet towards its prey.

indicate a close link with leopards and cheetahs–felines that likewise fail to recognise rigid laws relating to territorial privilege and frequently stray across the frontiers of one another's hunting grounds.

The lynxes of the high sierra

The only scientific information about the lynxes that live in the mountain regions of central Spain has been provided by the young naturalist Jesús Garzón Heydt, and the following facts are based on his findings, here published for the first time.

In the mountains of Castile the Spanish lynxes roam wild, barren, empty terrain as well as inaccessible slopes overgrown with heather and gorse and dotted with oaks and cork oaks. In most of this region, except for parts of the Pyrenees, they owe their survival to the still-abundant rabbit population, for these hardy mammals constitute the principal item in the carnivores' diet. Even here, however, myxomatosis has had its effect and many of the predators have been obliged to extend the bounds of their hunting territories.

In 1960, while the epidemic was at its peak, a number of lynxes were sighted and caught in areas where nobody had suspected they existed. Recent surveys suggest that now the rabbits appear to have gained the upper hand over the disease, the predators may be more gravely threatened for another reason. Reafforestation with rapidly growing coniferous trees is altering the whole environment. Traditional wild plants such as cistus and gorse have been rooted up and tall oaks put to the axe, leaving ugly bare patches, some of which have been replanted with pines. Many lynxes, abandoning regions which had formerly provided refuge or driven out by fires lit to burn dead vegetation, were killed; others just managed to survive.

Nevertheless, the traditional plants of the sierra are tough and cannot be completely destroyed. As they begin to grow again alongside the young conifers there is renewed hope that they may be allowed to develop sufficiently to save the remaining lynxes of these mountain regions. Yet it may only be a brief respite. The apparent policy of the forest authorities is to keep cutting back the undergrowth at regular intervals. Moreover, when the conifers reach a reasonable height their branches touch and interlock so as to form a permanent canopy which gives a dense shade, impeding the development of the new shoots of other species. Thus instead of the dense, tangled underbrush where rabbits once romped freely there are again flat, bare patches covered with slippery pine needles. Since there is no longer enough grass or low foliage, the rabbits tend to abandon the pine woods for fresher pastures elsewhere and the predatory lynxes are once more deprived of their principal food.

In some regions it is not too late to adopt sensible policies which would prevent such a situation arising. All that has to be done is to earmark certain parts of ıe maquis for the planting of oaks alongside conifers. This combination, apart from allowing both predators and prey to live together in ideal surroundings, would also encourage many insect-eating birds to nest. This

would be an important contribution to safeguarding the forest environment by re-establishing a healthy balance between flora and fauna.

The area most closely surveyed by the Spanish naturalist Heydt was a wild mountainous region of some 10,000 acres in Estremadura. Over a period of several years he followed the movements of a dozen lynxes. By direct observation and examination of their footprints as they roamed to and fro he charted the individual portions of territory belonging to each animal (as shown in the map which appears on page 15).

This naturalist discovered that although the various territories were established on different types of terrain, they all tended to cover about 750 acres and were all provided with safe refuges that afforded concealment from enemies and prey alike. They contained hiding places where the predators could lie in ambush and were well provided with the type of vegetation that would attract their habitual prey. The lairs were sometimes situated on an empty stretch of rocky ground with the occasional isolated oak and clumps of ferns and gorse (A,D,G,H,I,J,); alternatively, they might be established in thickets of bramble, gorse and arbutus (C,K,L), in pine forests with a dense cover of undergrowth in the form of heather or cistus (B,E), or in a forest of oaks (F).

At one time the Spanish lynx preyed principally on rabbits, which made up more than half of its daily food supply. After the myxomatosis epidemic, however, the predator was obliged to modify its feeding habits, extending its range to include a larger number of rodents, birds and domestic animals.

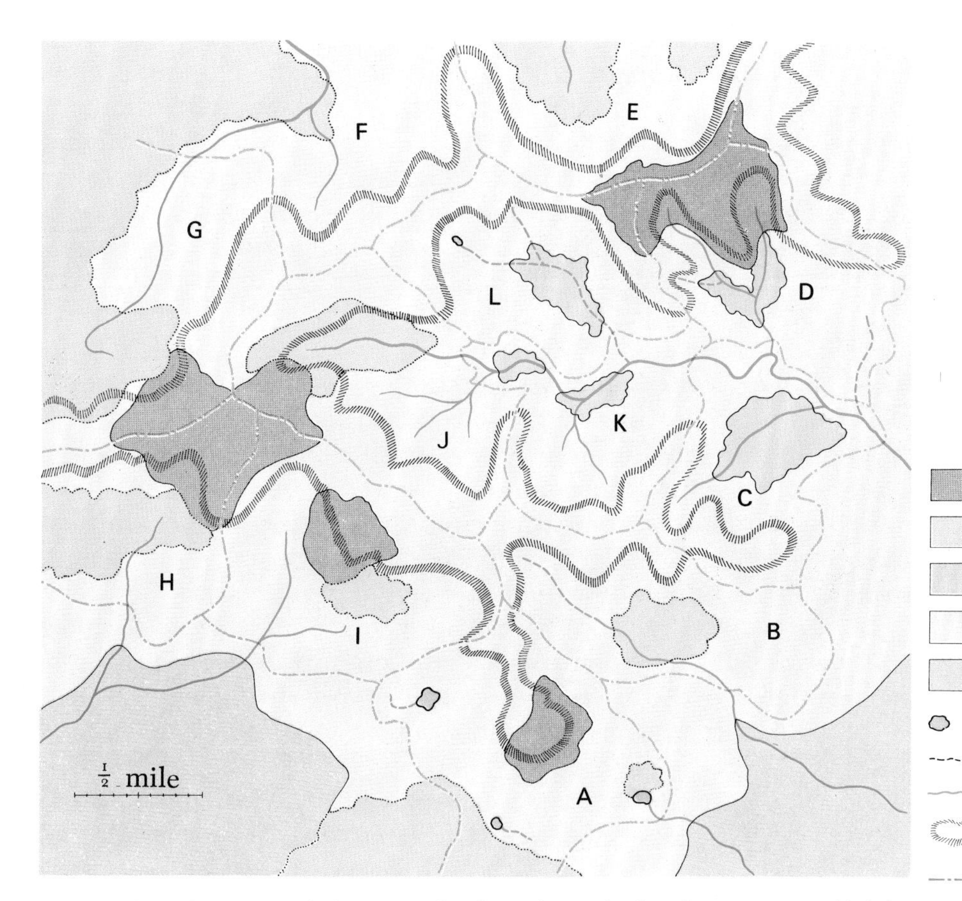

Map prepared by Jesús Garzón Heydt showing an area of Estremadura where he studied the territorial behaviour of twelve Spanish lynxes (A–L). The topographical key appears below.

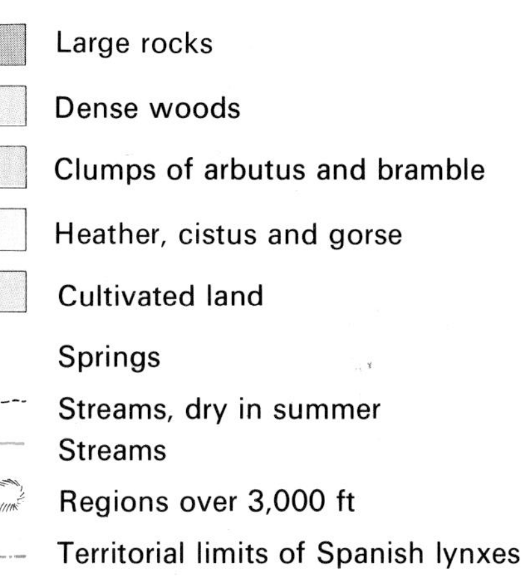

Another feature of the map is that there is fresh water available in every section of territory, regular drinking being of prime importance in the lynxes' daily activities. Here it is found either in the form of a stream or a spring, some of which do not run dry, even in severe periods of drought.

Thanks to footprints left by the occupant of the territory marked L on the banks of the only spring rising in that particular area, Heydt showed conclusively that this lynx, like the others, was in the habit of visiting the same spot every night to quench its thirst. In August the pattern of prints indicated that the elderly female to whom this territory belonged was accompanied by two youngsters.

The most important section of each piece of territory is without doubt the hunting precinct, and the extent of this is solely determined by the amount of prey found therein at any given time. In years when food is in short supply this hunting area is enlarged. In order to do this the lynx drives out weaker rivals. The latter, compelled to roam far and wide in search of a new home, are likely to be tormented by hunger and the chances are that they will die. Those that are strong enough to survive may settle down in areas where the species had never formerly lived or in regions from which it had long since vanished.

The hunting grounds of the Spanish lynx in this part of the country are characterised by fairly level terrain, uninterrupted by hills and rocks, and well covered with clumps of oaks and thick undergrowth. The area teems with small mammals such as rabbits, fieldmice, voles and dormice and with a surprisingly large number of wild boars. Although the lagomorphs and rodents are regularly hunted, it has always been assumed that

The Spanish lynx looks much like a large domestic cat. Characteristic features are the spotted coat, short tail, long tufts of hair sprouting from the ears and thick white cheek ruffs which give the face a squarish outline.

lynxes did their utmost to avoid clashing with the dangerous boars. Yet Heydt came across the remains of the latter at several different times of the year along paths regularly frequented by one aged lynx, indicating that at least this predator had no qualms about attacking an adult boar weighing more than 200 lb, in addition to the more vulnerable younger animals.

The explanation of this unaccustomed behaviour may be that after the disappearance of their principal enemy, the wolf, from Estremadura, the wild boar population increased rapidly. Starving lynxes, frustrated by their inability to catch enough rabbits because of the myxomatosis epidemic, would then have modified their hunting patterns to meet the critical situation. In the course of so doing, they would hardly have turned down the chance of acquiring this valuable source of protein. After a time it is possible that they became quite adept in capturing boars and that they performed a service to farmers by reducing the numbers of these animals which did much damage to crops.

In the central mountain regions the lynxes appear to mate about a month later than those of the maquis and swamps. The earliest date for the animals to couple is mid-February.

An elderly mountain peasant, familiar with the comings and goings of lynxes, claimed that many years previously he had seen six males in the company of a single female during the breeding season, suggesting a marked departure from the normal male-female relationship. Although based on distant memory and unconfirmed by other witnesses, this kind of report is intriguing for naturalists and should not be dismissed out of hand. It may contain sufficient truth to stimulate further investigation on this interesting subject.

Undoubtedly the carnivores do abandon their habitual territory when the time comes for them to seek a mate. Were they not to roam abroad, many sexually mature animals would not have the opportunity to procreate and clearly the future of the species would be in jeopardy.

It is estimated that there are nowadays some 150 pairs of Spanish lynxes roaming the Guadalquivir delta, the Sierra Morena and the mountains of Toledo and Estremadura, with a few stray individuals to be found in the mountains near Tortosa as well as the highlands of Portugal. Nevertheless, this tally may be too optimistic. When one bears in mind that at least 30–50 lynxes are caught and killed every year, that the destruction of traditional habitats remains unchecked, and that a number of the carnivores die after feeding on poisoned bait, it is evident that this beautiful animal is still in serious danger.

Apart from the Iberian peninsula, there are not many places where this threatened species still runs wild. Only a few are today seen in the Caucasus, the Balkans, the southern Carpathians, Greece, Albania and Macedonia. In other parts of Europe where they formerly ranged – Italy, Sicily, Sardinia and southern France – Spanish lynxes have completely disappeared.

It is fervently to be hoped that conservationists and governments will collaborate in making sure that this handsomely spotted carnivore – the only lynx whose name is listed in the IUCN's Red Data Book – is saved.

Canada lynx (*Felis canadensis*)

Spanish lynx (*Felis pardellus*)

European lynx (*Felis lynx*)

Bobcat or bay lynx (*Felis rufa*)

Caracal (*Caracal caracal*)

Of the five surviving lynx species, only the Spanish lynx is in imminent danger of extinction, its name appearing in the Red Data Book on Endangered Species, published by the International Union for Conservation of Nature and Natural Resources.

Destroying the forests

Man's need for raw materials has led him to exploit the earth's natural resources and scientists are now issuing warnings that if this continues in the present indiscriminate fashion, we are likely to face a major crisis in the not-distant future. Our mistake is to assume that what we take from the earth can automatically be replaced and that nature is inexhaustible.

Deforestation is a case in point. In Europe the destruction of trees in the Mediterranean basin began more than two thousand years ago and gained momentum during the Middle Ages to satisfy the demands of farming and to furnish timber, not only for building towns but also for equipping the navies of the great powers bent on overseas conquest.

It was not until the 16th century that an alarm call went up as the demand for timber began to exceed available supplies. In this century and the next there were only spasmodic attempts at reafforestation. Later efforts too were only partially successful and some areas had by now been transformed into irredeemable wasteland. Elsewhere, as we have seen, bids to re-establish woods and forests by utilising only fast-growing tree species such as conifers could not prevent soil erosion, with disastrous consequences for wildlife.

CHAPTER 2

Aristocrats of the Mediterranean skies

Because birds of prey are so mobile, it is not easy to tie them down to any particular habitat. Although some are typical of maquis or steppe, others are equally at home in a variety of contrasted environments. Where the natural scene and climate are diversified, the distribution is even more difficult to determine. Thus in the Mediterranean basin, where forests are interspersed with stretches of open ground and where a period of drought is often followed by a bout of wet weather, representatives of almost all the characteristic raptor species of the Palearctic region are to be seen at various times of the year.

Hunters of the dense woodlands, such as the goshawk and sparrowhawk, take advantage of the beneficent climate and frequently nest in the Mediterranean during the winter. With their long tails and short, rounded wings, such birds manoeuvre with great skill and the mixture of trees and open terrain is ideal for their hunting activities. The peregrine falcon too, roosting on high rocks that overlook fertile valleys, swoops down on the level plains, unencumbered by trees or shrubs. But none of these birds is a typical raptor of the Mediterranean maquis, any more than Bonelli's eagle, which is normally found perching on rocky hillsides and mountain slopes along the coast and whose range extends eastwards into Asia.

The true maquis, with its patchwork pattern of copses, undergrowth and plain—sometimes perfectly flat, sometimes gently undulating—is ideal for birds of prey that have no need of specialised hunting techniques and are able to adapt to different conditions. The entire region is populated by a wide range of small and medium-sized animals which provide plenty of food for these aerial predators whose instinctive rule—like that of all hunters—is to expend the minimum amount of energy at all times.

Facing page : The short-toed eagle is a Mediterranean raptor which specialises in the capture of snakes, particularly colubrids. In this picture a male is seen bringing a half-swallowed prey to his mate in the nest.

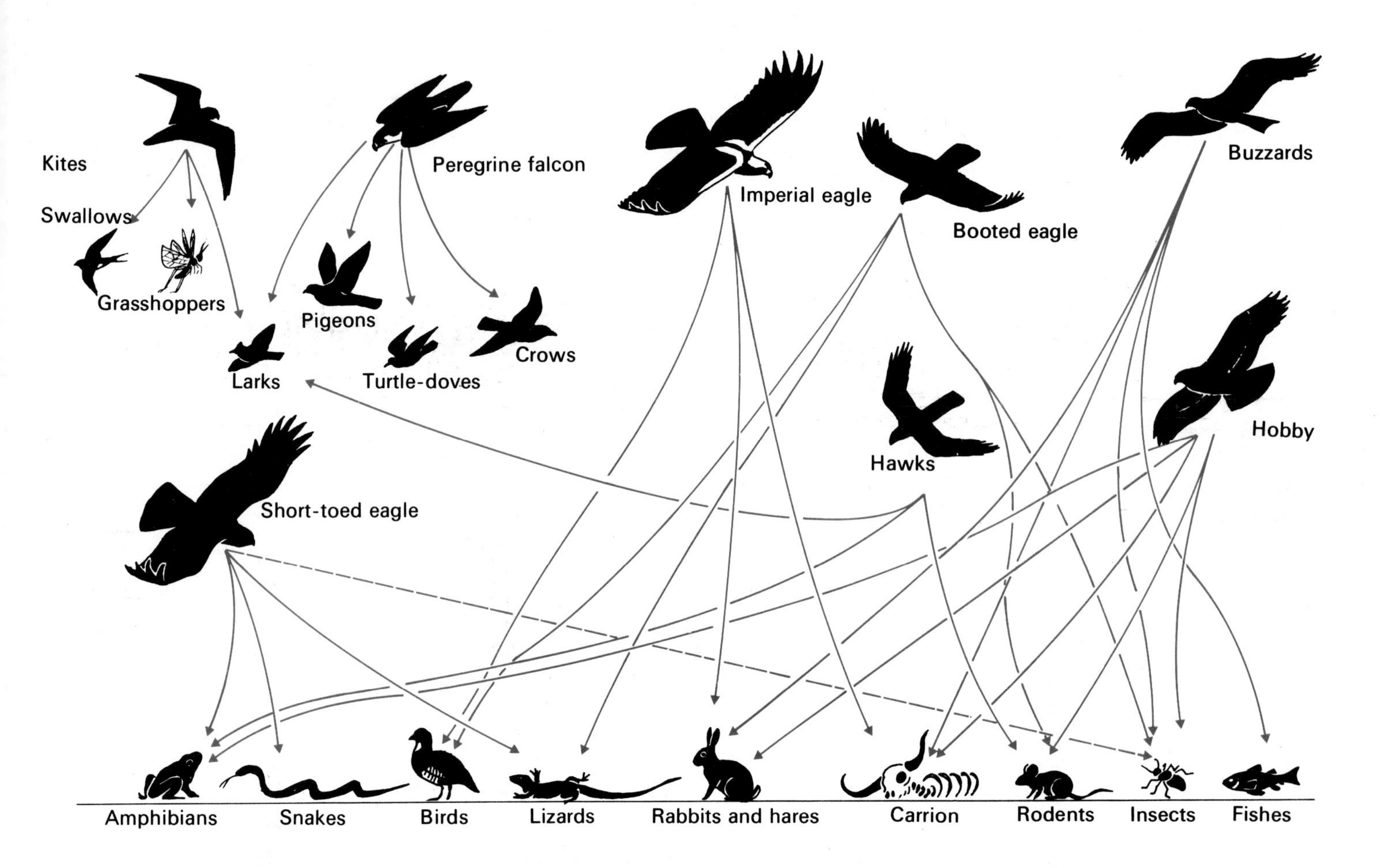

Because of its mild climate, varied vegetation and abundance of wildlife, the Mediterranean maquis attracts many birds of prey, some sedentary, some migratory. Most of them are unspecialised hunters with a wide range of victims, as can be seen from this diagram.

The tendency of these raptors is to wait patiently, hour after hour, until a suitable victim comes within range, rather than fly off in immediate pursuit of the first animal that happens to catch the eye.

Among such opportunistic hunters are the imperial eagle, the booted eagle–above all–the kites, all of which have a broad range of prey including rodents, hares and rabbits, lizards, birds, fishes, insects and carrion. A characteristic bird of prey of the maquis with more specialised habits is the short-toed eagle (*Circaetus gallicus*), which feeds mainly, but not exclusively, on snakes.

Surprisingly few European ecologists and ornithologists have devoted much time to studying the food habits of the continent's birds of prey. One exception is the naturalist Dr Castroviejo, who has published a number of works on this fascinating subject. He has pointed out that the customary diet of European raptors becomes increasingly diversified as they range southwards towards the Mediterranean basin. This observation is in fact valid not only for the Palearctic region but equally for other parts of the world, and is in keeping with well attested scientific findings elsewhere. In warmer countries, although there are a greater number of animal species, there are appreciably fewer individuals to each community. In such areas carnivores tend to specialise in certain types of prey. Once they have exhausted their food resources by concentrating their attacks on particular species (thereby reducing the numbers), the predators are doomed. Odds are more in favour of the opportunists–those hunters that enjoy a varied diet, adapt their methods to changing circumstances and consequently survive.

The detailed surveys conducted by Castroviejo have provided irrefutable proof that the birds of prey of the Iberian peninsula feed on amphibians (frogs and toads) and reptiles (snakes and lizards) to a much greater extent than do the raptors of central Europe. In the Mediterranean region the more abundant poikilotherms (animals with a body temperature varying in response to the outside temperature) play a much more important role in the food chain than do the homoiotherms (warm-blooded animals). Thus the eyed lizard, Montpellier snake and viperine snake are the counterparts–particularly in the Cantabrian Mountains–of the voles and other small rodents which constitute the basic food of the falcons of the eastern European plains.

Apart from the bird-eating peregrine falcon (of which some dozen pairs have been sighted in central Spain during the past five years), and the Eleanora's falcon (which appears to reject snakes altogether), all the other familiar birds of prey found in Spain include amphibians and reptiles in their diet, whether they consume them alive or dead. Even those winged hunters that specialise in catching mammals and pigeons–such as the goshawk–have clearly modified their habits according to opportunity, as is proved by the discovery of the remains of eyed lizards in goshawks' nests in different parts of Spain, notably the Sierra de Gata, the Obarenes Mountains and the highlands of Toledo.

Geographical distribution of the imperial eagle.

The rare imperial eagle

In the early part of spring, as the days gradually get longer and warmer, the swamplands of the Guadalquivir appear in all their natural splendour. After the numb, bleak winter the whole region seems to explode with life. The last remaining flocks of migrating birds prepare to take off on the return journey northwards to their breeding grounds. Already the moorhens and ducks are occupied with their broods among the reeds. Grebes, magnificently attired in their nuptial plumage, swim in the marshes which are brightly dotted with buttercups. In the branches of tall oaks herons are noisily engaged in building their nests. Wild sows grunt their way through the heather, marshalling their squealing piglets to safety; and from time to time a Spanish lynx, belly close to the ground, slips silently through the clumps of rock-roses.

All these comings and goings are attentively watched by a huge bird of prey perched on the top of a pine. At dusk the majestic outline of the imperial eagle (*Aquila heliaca*) is etched against the fiery red sky as it surveys its hunting grounds below.

There is no mistaking this fine bird for there are several features clearly distinguishing it from all related species in the Holarctic region. In size it may be compared with the golden eagle, but it differs from the latter in having a large, pale yellow patch extending from above the eyes and down the top of the neck as far as the shoulders and front of the wings, where it merges with more conspicuous areas of pure white.

This winged sentinel of the Doñana reserve is certainly the rarest and most vulnerable of all European birds of prey. Although it is still seen in the skies over the southern part of the Iberian

IMPERIAL EAGLE
(Aquila heliaca)

Class: Aves
Order: Falconiformes
Family: Accipitridae
Length: 31½–33½ inches (79–84 cm)
Wing-length: 22¾–26½ inches (57–66·5 cm)
Wingspan: 76–84 inches (190–210 cm)
Weight: 6–8 lb (2·5–3·5 kg)
Diet: small and medium-sized mammals, birds, reptiles, carrion
Number of eggs: 1–3, usually 2
Incubation: 43 days

Adults
Large bird with blackish-brown plumage. Neck and top of head pale yellow; shoulders adorned with white patches. Tail has 5–7 grey bars across it. Tail feathers are shorter and squarer than those of golden eagle. Cere and feet yellow.

Young
Fledglings are covered with white down. Later the young are variable in colour, from yellowish-brown to uniform reddish-brown. Tail darker than that of adult. Pale marks are visible in the centre of either wing.

Golden eagle
(*Aquila chrysaetos*)

Imperial eagle
(*Aquila heliaca*)

The imperial eagle is equipped with shorter and less powerful talons than the golden eagle. Thus, unlike its relative, it does not attack young deer or goats, concentrating on smaller prey such as rodents and birds.

Facing page : Imperial eagles build their huge nests at the top of a tall tree and the adjoining branches provide excellent vantage points for surveying the surrounding country. The species may be distinguished from the golden eagle by the pure white patches on the shoulders.

peninsula, such visits are becoming ever more infrequent. There are too many people who, with little regard for beauty and complete ignorance of biological laws, simply treat it as a harmful species and reach for their guns whenever it appears. Little wonder that the splendid imperial eagle is threatened with speedy extinction unless urgent action is taken to save it.

There are two races of imperial eagle, one in Spain and the other in eastern Europe and Asia, the range of the latter including Greece, southern parts of the Soviet Union and certain regions in central Asia, Siberia and Outer Mongolia. The Spanish race (*Aquila heliaca adalberti*) may be distinguished from the eastern race (*Aquila heliaca heliaca*) by the unspotted plumage adorning the shoulders of the adult bird.

In former years the imperial eagle could be found throughout Spain and was believed to nest in oak and pine forests, including those of High Castile. Unfortunately the population has dwindled alarmingly in the past thirty years. Around 1940 the raptor still roosted only a few miles outside Madrid, on El Pardo mountain. Since then it has never been sighted there and in fact the Doñana reserve is the only place where it is now found in any numbers. Elsewhere in Spain, a few isolated pairs have succeeded in surviving in the Sierra Morena and the mountains of Estremadura and Toledo. The total population in the country is reckoned to be only thirty pairs.

It seems incredible that one of the most imperilled bird species in Europe, indeed in the world, should continue to be at the mercy of poachers, intent only on possessing a stuffed specimen under the feeble pretext that by so doing they are performing a service to the community by protecting local game.

Death from the sky

In contrast to the golden eagle, which prefers to choose a vantage point on a steep mountain ledge, the imperial eagle generally takes up a commanding position overlooking flat country where there are only gentle hills and scattered woods and copses. In the Guadalquivir swamps – the only place in the world where sizeable colonies are still found – the raptor is most frequently seen on plains dotted with occasional trees, usually oak or pine, and areas covered with cistus and heather. Flying lazily over moors and marshes, the predator keeps sharp watch for any sign of activity below, particularly a scampering rabbit, always a coveted victim.

In the course of these reconnaissance flights, different pairs of eagles follow criss-cross paths so that it is difficult to distinguish the boundaries of individual territories – imprecise at the best of times. In one district covering some 40,000 acres, ornithologists reported sighting seven nests, suggesting that the hunting grounds of a single pair of these birds might extend to more than 5,000 acres.

As they fly back and forth, each male and female can cover a wide area and yet subject every square foot of terrain to the closest scrutiny. When there is a good deal of animal activity on the ground the birds are not obliged to travel such great

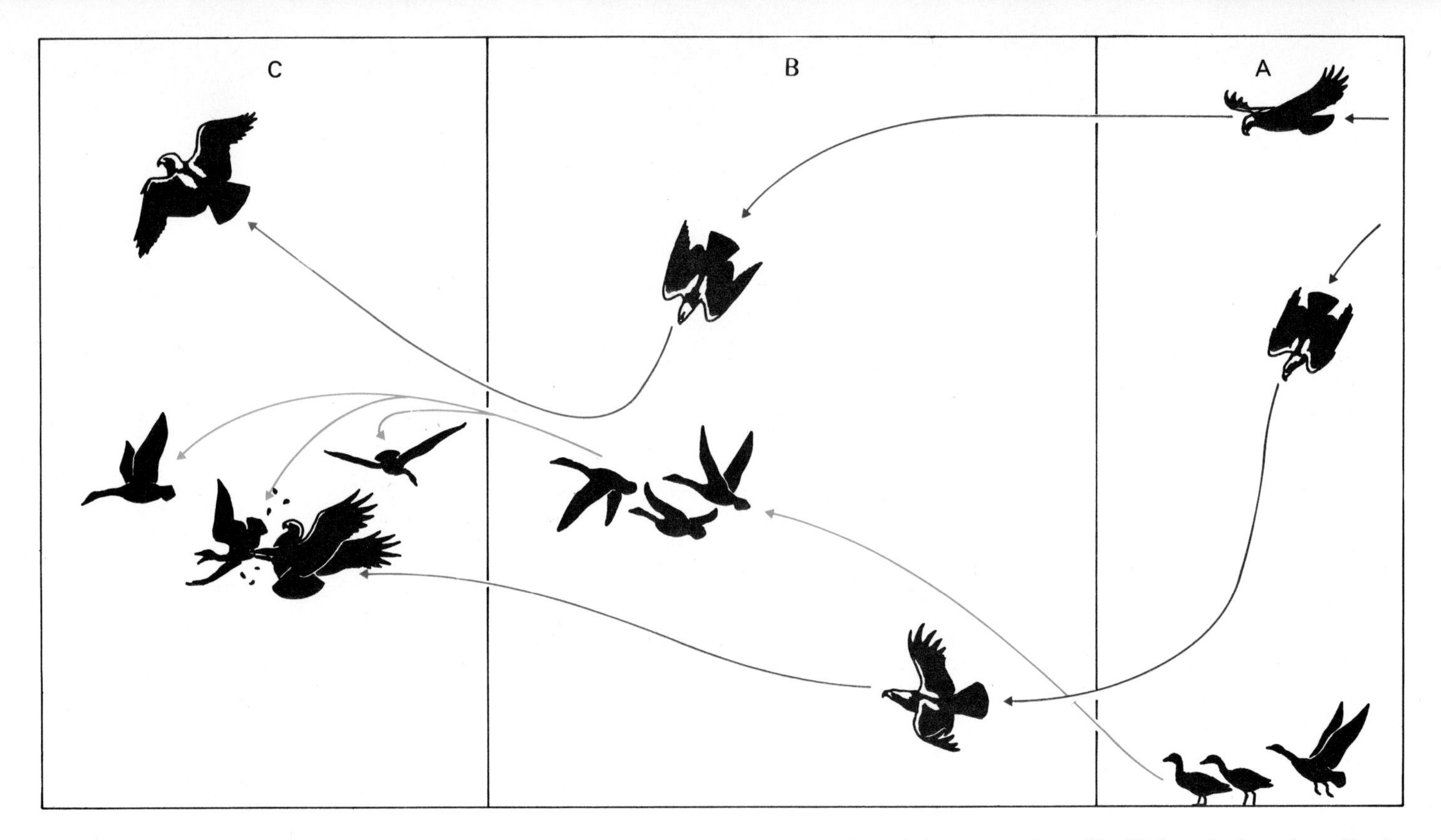

Although imperial eagles usually hunt alone, they sometimes form pairs for a joint operation. Dr Valverde has described the manner in which two eagles will attack a flock of wild geese. Flying at the same altitude, the raptors separate, one going into a sharp dive (A), the other intercepting the alarmed geese (B) and driving them into the path of the first bird (C).

The imperial eagle has various methods of capturing prey on the ground, either perching on the branch of a tree or spiralling up into the sky on warm thermal currents, which affords it a broad field of vision. As soon as it sights its victim–often a rabbit–it swoops down at an oblique angle, skilfully avoiding any natural obstacles that are in the way.

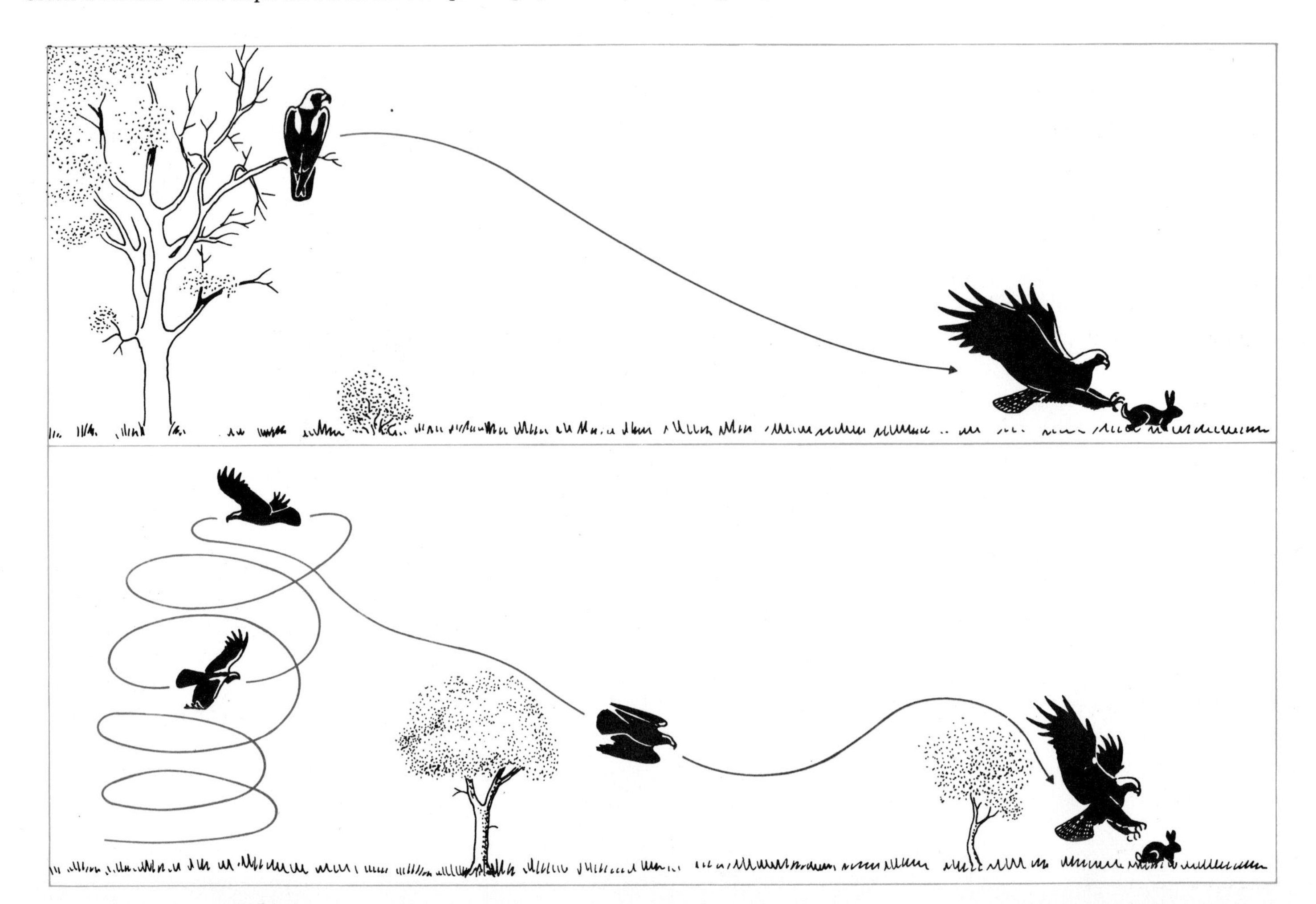

distances, nor to make such a detailed and systematic search for prey. All they have to do is to circle over a particular spot and wait for an unsuspecting victim to appear, finally launching an attack. If it is an average-sized animal such as a pigeon or rabbit, the eagle will carry it off to a tree or rock and devour it at leisure. Larger, heavier prey is usually consumed on the spot. The maximum weight of any animal carried back to the eyrie is 5–6 lb.

The imperial eagle has two principal hunting techniques. Either it perches on a tree or outcrop and swoops like lightning on its prey, or it takes to the air to chase other birds and catch them on the wing. Thus in the morning it will sit on a branch situated close to a site where its habitual prey is likely to be found. Later in the day it may extend its attacks to other animals, either ambushing them on the ground or chasing them in the air. By evening the predator will once more be keeping watch on fields where rabbits can be expected to be outside their burrows feeding.

The imperial eagle is a great opportunist at all times, often swooping on a bird which might be imagined to be too tiny to merit attention, or diving to impale a fish basking in shallow water. In the marshes two eagles may mount a joint operation to catch wild geese. One bird flies over the flock to alarm and distract them, the other coming up from the rear to pick off a straggler. But there is a generous choice of prey in these parts, including rabbits, waders, moorhens, gallinules, ducks, spoonbills, flamingos and other familiar species of lake and pond. In addition there are large lizards, cicadas and other insects.

The eastern race of imperial eagle preys principally on ground squirrels. Indeed, so dependent are they on these rodents that in seasons when the latter fail to breed as prolifically as usual, the raptors lay fewer eggs, with the obvious result that their numbers decline more rapidly than would otherwise be the case.

Spring and summer

Spring is the season when the imperial eagles mate, prior to which each male and female performs dramatic aerial manoeuvres. For hours on end the two birds soar through space, signalling to each other with clamorous cries. They swoop and almost collide head-on in affectionate, mock combat; then they both point for the ground and plummet down in a dizzy, spiralling double dive. As they straighten out and resume their horizontal flight pattern, the bird in the lower position will sometimes flop over on its back so that it is literally flying upside-down. For a few seconds they link claws, still wheeling round in circles and emitting their loud mating calls; then they let go and after a further sequence of acrobatic swirls come in to land, doubtless exhausted, on a tree. Shortly afterwards they couple.

In Europe the female lays her eggs between mid-February and the end of March. The eggs, usually two, deposited in the nest within 48 hours of each other, are whitish with grey or purple flecks. In contrast to the behaviour of many other eagles, the male not only hunts for food but also takes his turn, at intervals, to

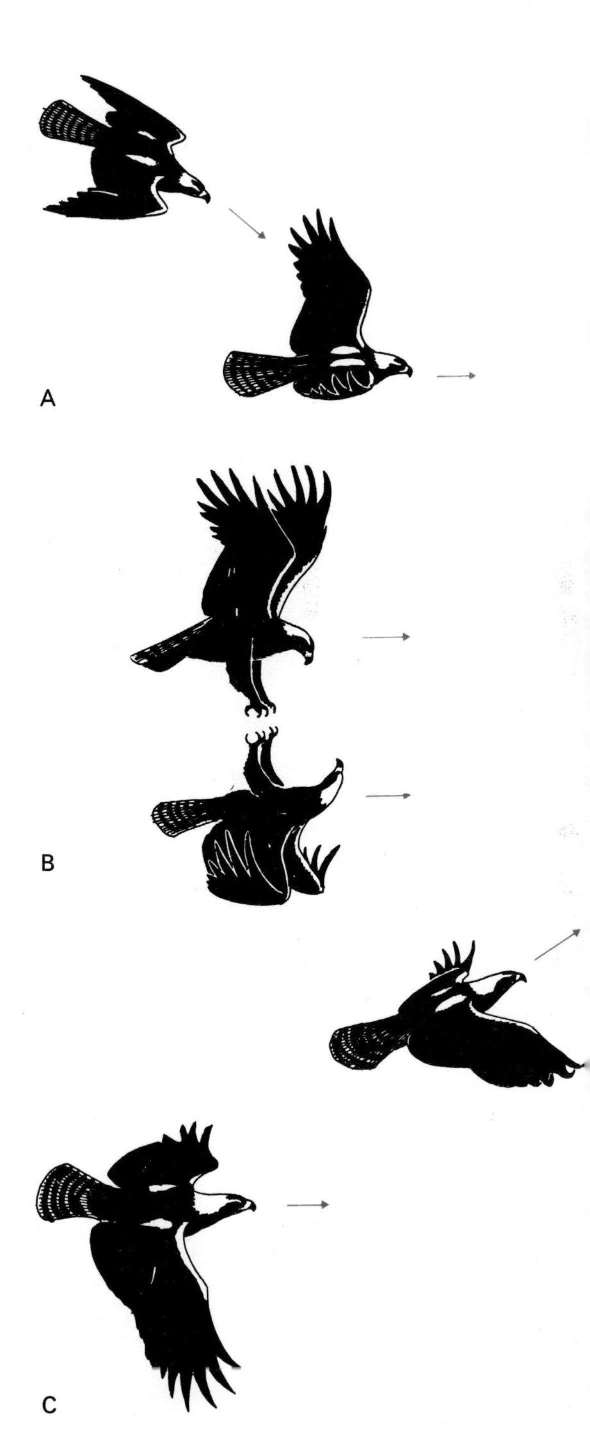

Imperial eagles perform amazing aerial acrobatics in the course of their nuptial displays. The most spectacular manoeuvre begins with the male pursuing the female, who then turns suddenly on her back to grasp the claws of her partner. The two birds fly together in this position for a few moments, emitting sharp cries, then separate.

An imperial eagle's brood generally comprises two fledglings, which are attentively reared by both parents. As the pictures on this and the facing page show, the newly born eaglets are covered with pure white down, but within a couple of weeks have developed brown feathers, easily distinguishable from the almost black plumage of the adults. They are now old enough to embark on their first flights but continue to be fed by the parents and to spend the night in the eyrie.

incubate the egg. After 43 days the first pure white fledgling is hatched, then two days later the second, this time-lag explaining why one eaglet is normally smaller than its sibling.

Two weeks after birth, the stems of the breast feathers appear. Growth is then fairly rapid and within eight weeks the white down has almost disappeared. Life in the eyrie is more peaceful than in that of the golden eagle (where the stronger eaglet kills the weaker one) and both youngsters develop side by side, jealously guarded and supervised by the mother. She spends long periods with them in the nest and when the male arrives with prey, divides it up into equal portions for her offspring.

After about six weeks the female also ventures out to help her mate find food for the greedy eaglets and within another fortnight these are ready to embark upon their first flight. They are prudent enough, however, not to abandon the eyrie just yet and spend the intervening time sleeping and awaiting the return of the parents, who continue to feed them. Eventually the eaglets spend more and more time away from the parental nest, yet still come back to it at night.

Once they are old enough to fly they are old enough to hunt, and before the arrival of autumn they are learning to fend for themselves. Their plumage is by now pale brown, so that they can be distinguished very easily from the adults with their dark brown livery, yellow head and clear white shoulder patches. Not until they are five or six years old will the young birds take on the characteristic colours of the adults and as they grow they adopt a wide range of intermediate hues, the brown feathers gradually

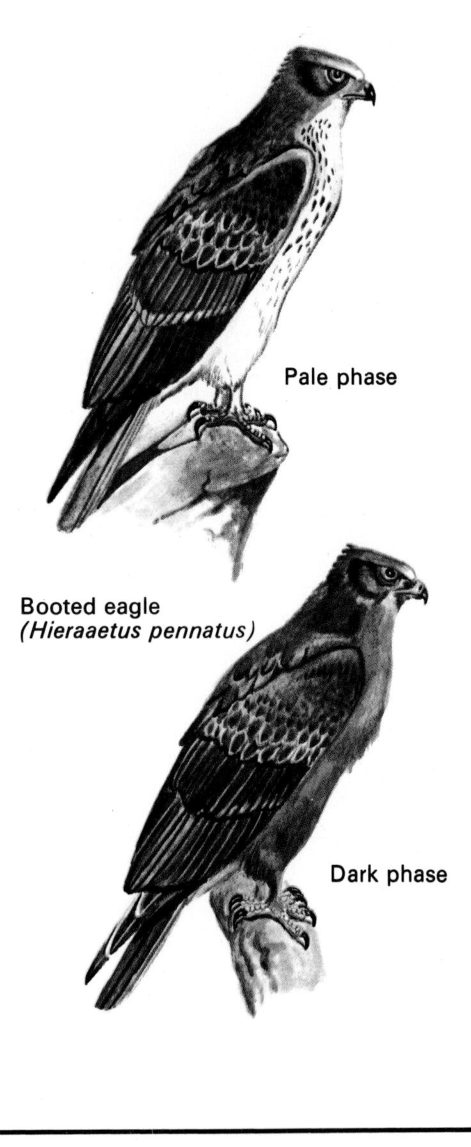

Booted eagle
(Hieraaetus pennatus)

becoming darker.

As a general rule, only those eagles that have acquired the distinctive dark brown plumage will mate. Yet if for any reason one of the mature birds making up a breeding pair fails to appear, its place may be taken by a younger, immature individual which will then mate with the remaining adult.

The diminutive booted eagle

As the first gusts of the south wind herald the spring and the oak buds burst open, the lovely little booted eagle (*Hieraaetus pennatus*) returns to the woods. So called after the boot-like tufts of feathers that cover the legs from tarsi to toes, this bird is the most delicate and arguably the most beautiful of the entire eagle family.

At the end of March and the early part of April the sky resounds to the strident yet harmonious calls of these delightful birds as they perform their complicated nuptial flights. Each morning they can be seen soaring up into the sky until they are almost lost to sight, then reappearing as they tear earthwards in a plummeting dive which levels out just before collision with the ground seems inevitable. Between dives they wheel in unison, sometimes linking claws, allowing themselves to be carried along at random by warm air currents.

These acrobatic displays take place more or less directly over the site where the birds will eventually build the nest. Normally this is constructed on the trunk or main branch of an oak or other large tree; but the place is not all that important and in fact any tree, shrub or natural cavity will suffice.

Both eagles help each other to build the nest, patiently flying to and fro carrying thin branches and piling them up to form an enormous framework which may be well over a foot in diameter. This will be carefully lined with green twigs and leaves which happen to be available close by.

Around mid-April it is usual to find two white eggs in the nest (sometimes, though rarely, the number will be one or three). They are rounded, with rough shells. The female incubates them while the male stays nearby to repel other birds of prey that may infringe on the territory. Incubation lasts approximately 30 days and the fledglings hatch within 48 hours of each other. As in the case of the imperial eagle, there is a significant size discrepancy and in this species it is generally only the elder eaglet which survives.

When they first come into the world, the fledglings weigh about $\frac{1}{4}$ lb and receive large quantities of food from both parents, mainly in the form of eyed lizards, choughs, blackbirds, shrikes and crested larks, with the occasional turtle-dove, pigeon or young partridge. The adults employ a variety of hunting techniques to capture their prey, diving almost vertically at high speed, weaving an expert course in and out of woods—narrowly avoiding the closely packed tree trunks—and even swooping down to dislodge birds from low branches and to flush out lizards from the undergrowth. They literally cram their offspring with bits and pieces of prey, themselves consuming those parts that are too

BOOTED EAGLE
(Hieraaetus pennatus)

Class: Aves
Order: Falconiformes
Family: Accipitridae
Length: $18\frac{1}{4}$–$21\frac{1}{4}$ inches (46–53 cm)
Wing-length: 14–$16\frac{1}{4}$ inches (35·2–40·3 cm)
Wingspan: 44–$52\frac{3}{4}$ inches (110–132 cm)
Weight: female $1\frac{3}{4}$ lb (700 g)
male 2 lb (900 g)
Diet: small and medium-sized mammals and birds, large lizards
Number of eggs: 2 (exceptionally 1 or 3)
Incubation: about 30 days in wild, 39 in captivity

Adults
About the size of a buzzard but with a longer tail. Plumage shows variations unrelated either to sex or age. The majority are pale, the remainder dark, the difference being in the colour of the lower parts—white in the former, dark brown in the latter. The rest of the plumage is the same in both phases—forehead white; head and top of neck yellow with chestnut stripes; back, wing coverts and remiges brown to black; rectrices dark brown above, cream below. Cere and feet yellow.

Young
Fledglings covered with white down. Developing bird much resembles pale phase of adult.

tough for the youngsters' tender beaks. Thus the heads and feet of birds and lizards are torn off and only the softer, more easily digestable portions of the body fed to the eaglets. In this way every part of the victim serves as food, either for adult or young.

Gorged with food, the eaglets grow rapidly. Within ten days their feathers begin to appear and after five weeks they are capable of independent flight and ready for instruction in hunting methods. They begin by keeping their parents company in the daily quest for food and soon acquire all the adult skills.

At the onset of autumn, as days become shorter and the first frosts harden the ground, the booted eagles–which until then have spent each night sheltering in their tree–strike out for the south. Slowly and methodically, alone or in pairs, the birds cross the sea by way of the Straits of Gibraltar and head for winter quarters–thought to be between Senegal and Cameroon.

Much confusion has arisen over identification of this species as a result of the very variable colours of the plumage. In fact there are two clearly contrasted colour phases–pale and dark. But even with this distinction positive identification is complicated by the fact that no two individuals are alike. Some are almost pure white, others chocolate-brown, and between these extremes there are examples of every imaginable intermediate shade. All that can be said is that the pale phase seems to be more common than the dark phase. Dr Valverde estimates that for every dark booted eagle there are seven pale ones. Paul Géroudet reckons that out of every ten eagles, eight will be pale and that the dark ones will mate with each other.

Compared with other members of the family, the booted eagle is virtually a dwarf, being roughly the same size as a buzzard. Yet this is offset by greater aggressiveness and an uncommonly versatile flight pattern which includes more spectacular dives.

In addition to Spain and Portugal, the booted eagle nests in North Africa, eastern Europe and southern Asia; and in winter it migrates as far as India as well as Africa. The subspecies *Hieraaetus pennatus pennatus* is found in Spain, North Africa and through Europe to the Caucasus; the more numerous *Hieraaetus pennatus haterti* is an inhabitant of eastern Asia.

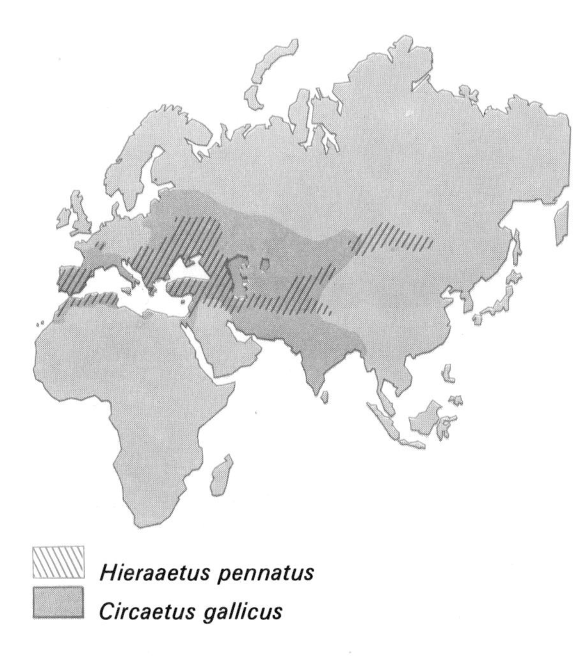

Geographical distribution of booted eagle (*Hieraaetus pennatus*) and short-toed eagle (*Circaetus gallicus*).

The short-toed eagle

The courtship display of the short-toed eagle (*Circaetus gallicus*) is an uncommonly graceful and beautiful spectacle. Male and female fly along together, giving out soft, gentle calls, their paths continually crossing and intertwining. At one moment they are no more than two distant specks in the sky, then a few seconds later their outlines take clear shape as the birds dive to graze the tops of the pine woods. As with other raptors, the birds bring the display to a climax by flying parallel to each other, one of them turning upside-down to face its partner above, the two pairs of claws meeting and touching for a brief moment. All this ostentatious activity is confined strictly to the breeding season, for at all other times the birds' behaviour is restrained and unspectacular.

SHORT-TOED EAGLE
(Circaetus gallicus)

Class: Aves
Order: Falconiformes
Family: Accipitridae
Length: $25\frac{1}{4}$–$27\frac{1}{4}$ inches (63–68 cm)
Length of tail: $11\frac{1}{2}$–$13\frac{1}{4}$ inches (28·7–33 cm)
Wing-length: male $20\frac{1}{2}$–$22\frac{1}{2}$ inches (51–56·5 cm)
female $20\frac{3}{4}$–$24\frac{1}{4}$ inches (52–60·5 cm)
Wingspan: 64–72 inches (160–180 cm)
Weight: $3\frac{1}{2}$–$4\frac{1}{2}$ lb (1·5–2·1 kg)·
Diet: mainly large reptiles, especially snakes and lizards; also some mammals and, occasionally, medium-sized birds
Number of eggs: one
Incubation: 47 days

Adults
Upper parts brown or greyish-brown; lower parts, underside of wings and rectrices almost white. Throat and top part of breast light brown or striped chestnut. Rounded head, small bill, large orange eyes. Long tail, the upper part with three or four black bands. Cere and feet grey.

Young
Fledglings covered with white down. Later the lower parts of the body are the same colour or a little darker than those of the adult, with several black stripes.

The short-toed eagle feeds mainly on snakes and lizards and is often seen among rocks and on sandy beaches where such reptiles are frequently found basking in the sun.

Preceding pages : When it has caught a snake, the adult short-toed eagle carries it back to the nest and tears it up into convenient pieces for its hungry offspring. When the fledgling is a little older it does not wait for the parent to regurgitate the half-swallowed reptile but engages in a friendly tug-o'-war, using feet and bill to get its share.

Although sometimes known also as the snake eagle, this bird of prey is actually more akin to kites than eagles proper. The European subspecies is *Circaetus gallicus gallicus*, others being *Circaetus gallicus beaudouini*, living in the western part of tropical Africa, and *Circaetus gallicus pectoralis*, an inhabitant of North Africa.

In Europe the short-toed eagle arrives at its nesting zone at the beginning of April. The courtship display is followed by coupling and the construction of the nest, usually at the top of a tall tree more than 100 feet from the ground. Pines and oaks are equally suitable, both offering adequate concealment from enemies. The nest consists of a light frame of slender branches, about three feet in diameter, which is then lined with pine needles or green leaves.

The short-toed eagle occasionally builds an eyrie on a rock ledge or may even take over a nest previously occupied by another species. Male and female do the constructional work together, the former concentrating on the collection of materials, his mate carefully assembling them into the finished nest.

When the task is completed, around the end of April, the female is ready to lay one large, round, white egg which she then incubates uninterruptedly for about 47 days. Meanwhile she is spared any concern about food by her mate, who undertakes all responsibility for hunting and shares his prey with her.

The newly hatched fledgling is extremely puny and weak, weighing little more than an ounce. But it is soon swallowing small snakes and puts on weight rapidly. The first feathers appear on the breast and within 45 days cover the whole body.

At this age the young bird spends most of its time and energy learning to fly. The parents supervise these efforts and continue to bring the youngster plenty of food. Later they teach it to capture its own food by dangling a snake in front of its bill, then pulling it away and once more bringing it within range. The young bird soon realises that it is expected to fight for its meal and engages in a gentle struggle with the parent, finally succeeding in grabbing the dead reptile by the tail and tugging it free. Another important aspect of the youngster's development is the ability to conceal itself in the nest and not move a muscle in the event of an intruder approaching.

When it is about two and a half months old, the young short-toed eagle quits the eyrie, though remaining in the neighbourhood for another six weeks. When autumn comes, it flies south to warmer climes for the winter.

During this apprenticeship stage, the immature bird gradually acquires the skills required for later life. At first it glides clumsily, yet despite this awkwardness and the disproportion between its wingspan and the size of its body, its flight pattern already shows the rudiments of adult grace and elegance.

Because it is so light in weight, the short-toed eagle soars up with the greatest ease on rising thermal currents. Sometimes it perches on a rock to locate a potential victim, for with its keen eyesight it can immediately detect movement across open ground. As soon as it spots a snake (the favourite prey) it dives down from a height of 50–100 feet, trapping and immobilising the reptile between its short, curved talons, which are ideally shaped for this specialised style of hunting.

The raptor then breaks the skull of its victim by hammering it repeatedly with the bill until it expires. Sometimes the snake tries to bury its fangs in the feet of the predator but this seldom has any effect because of the thickness of the plumage protecting the front of the feet and the scales that cover the tarsi. If the snake is a small one it is swallowed whole, the entire process taking only a few seconds and often being completed in mid-air. Most victims are in fact of moderate size and quite harmless, such as the viperine snake and similar lowland species. Larger snakes measuring 6 feet or thereabouts are normally attacked only by a really strong individual. Venomous snakes are usually avoided altogether, even in places where they are plentiful and theoretically simple to catch, the reason being that the short-toed eagle is not immune to their poison. It is interesting to note that since a young bird has no way of knowing which snakes are potentially dangerous, it can only learn to distinguish one kind from another by watching its parents in action – a form of instruction which may last several weeks. It has been stated that this experience is passed on in the first place by an adult which has been bitten by a venomous species, but has survived. Tests on this point have been carried out on young birds removed from the eyrie, but in view of the fact that this creates an artificial situation the results of such experiments have to be treated with some caution.

In addition to snakes, the short-toed eagle feeds on lizards and rodents, but hardly ever attacks another bird. Examination of nest contents show that approximately 95 per cent of captured prey comprises snakes, 4 per cent lizards and the rest rodents. Adults sometimes include fish in their diet but this form of food

Bonelli's eagle
(Hieraaetus fasciatus)

Short-toed eagle
(Circaetus gallicus)

Despite its name, the short-toed eagle is not a true eagle. The latter, here represented by Bonelli's eagle, have feathers from tarsi to toes, whereas the short-toed eagle has a scaly covering to its feet, protecting it from the bites of the snakes which are favourite prey.

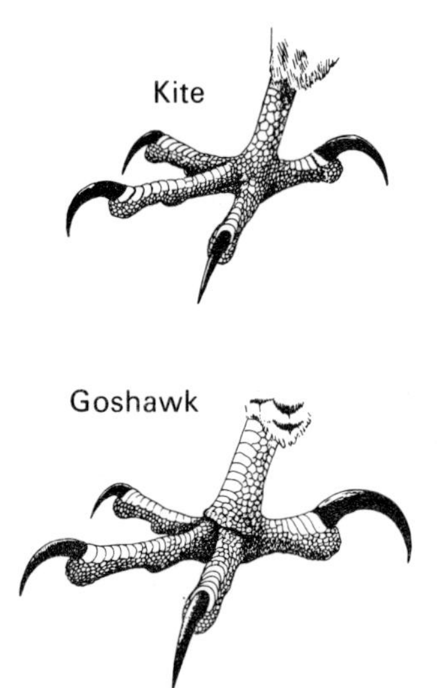

Most large birds of prey, such as the goshawk, have long, powerful talons to match; but the kites, despite their impressive size, have much shorter and weaker claws.

is rejected by young and immature birds.

Having undergone a long apprenticeship during the summer, the young eagle is ready to migrate when autumn comes, using those traditional flight paths which lead to winter quarters in East and West Africa.

The European subspecies is found in many southern and eastern parts of the continent, and its summer range extends to western Asia and North Africa. There are two principal migration routes. One passes over Gibraltar and Morocco into West Africa; the other crosses the region of the Suez Canal and takes the migrants into Eritrea and northern Sudan. Six months later they return northwards, flying singly and then forming pairs, heading for the ancestral nesting sites to rear a new generation.

Colour of plumage varies from one individual to another. The upper parts, the head and the throat are often chestnut, as are the irregular stripes which ornament the whitish areas on breast, belly and tail.

Kites of the Mediterranean

There are two species of genus *Milvus* in the Old World–the red kite (*Milvus milvus*) and the black kite (*Milvus migrans*). In the warmer countries the kite populations are sedentary, but those that nest in temperate regions undertake long migrations in spring and autumn.

Kites are sometimes mistaken for ospreys, buzzards or honey-buzzards, although there is a slight difference in size among these raptors. There is, however, one sure way of identifying the two kite species, for they are the only European birds of prey which have a forked tail. The red kite is the larger of the two, generally with paler plumage, and the division of its tail is more marked than that of its relative.

Red and black kites differ in other ways. Since they do not usually occupy the same habitats they display contrasted feeding and behaviour patterns. The red kite is frequently seen in forest clearings or in the pine and oak woods that are scattered through the Mediterranean region. Its food consists principally of reptiles, small mammals, fledglings (particularly those that are very young or physically incapacitated) and even insects. The black kite is basically a bird of the wetlands–swamps, marshes, lakes and quietly flowing rivers–and hunts rodents, fishes and amphibians. Both species sometimes feed on carrion.

The red kite

The red kite is a raptor of moderate size, about as large as a buzzard but with very much more elegant lines. The weight of the male varies from 1½ to 2 lb, that of the female from 2 to 2½ lb. The wings are long and narrow, the tints of the plumage rich and subtly blended, the tail deeply forked. The naturalist Paul Géroudet has likened this handsome bird to a brilliantly glowing giant swallow.

The red kite has an extensive distribution range which includes the greater part of Europe south of Scandinavia, down into the

Facing page : The handsome red kites, which help to maintain the natural equilibrium by feeding on sick animals and carrion are, like all raptors, gravely imperilled by the widespread use of pesticides. Since they are still freely shot as allegedly harmful birds, they are nowadays rare in most of their traditional habitats.

Geographical distribution of the red kite.

Mediterranean and Balkans and eastwards into Russia. It is also found in the Middle East, North-west Africa and the Canaries. The colonies living in the northern parts of the Palearctic region spend the winter in the north of India and Africa, but never cross the natural barrier formed by the Sahara Desert.

Whatever the country of origin, almost all red kites establish themselves in woodland, especially during the breeding season. Although they show a marked preference for the fringes of deciduous forests, they are also very common in the pine woods of the Mediterranean basin. The tree cover must not be too dense because the birds require broad stretches of open ground to exercise their hunting prowess, not being as expert in aerial twists and turns while hunting as are many other winged predators.

The red kite is not a particularly shy or retiring bird and is often seen in areas where there is considerable human activity. Deforestation–which so frequently proves catastrophic for some species–is no disadvantage for this hunter of the open spaces, and when fields are ploughed the bird is quick to swoop on any mice or beetles that are suddenly deprived of shelter.

A hunting foray generally takes the form of a series of probing, low-level flights over flat ground, in the course of which the bird may try to hover motionless in mid-air when it sights prey. To do this it turns into the wind, wings flapping strongly and tail spread transversely to the line of the body. At other times, although not ideally equipped for this type of hunting, the kite may soar to a greater height, circle above the woods and dive down on victims concealed in the thickets.

Despite the fact that the bird is not renowned for its manoeuvrability, it sometimes shows surprisingly quick reactions and is capable of a sudden burst of speed which enables it to chase and catch such swiftly moving species as pigeons and jays.

The red kite occasionally makes a raid on a chicken-coop or launches an attack against a hare or partridge (often to no avail). Such activities have not helped to enhance the bird's reputation with farmers and hunters, and although this harmful effect is much exaggerated, its bad name is sufficient to make it the target of sportsmen's guns all over Europe. As a result this attractive bird is fast vanishing from many of its traditional habitats and is seldom nowadays seen in numbers, except in Spain; but here, as elsewhere, the extensive use of chemical pesticides and their insidious effects on the food chain poses an additional threat to the raptor.

Mammals (especially rodents) and small birds figure prominently in the diet of the red kite; and in the Mediterranean countries reptiles–particularly Montpellier snakes, ladder snakes and eyed lizards–are common victims. Amphibians and large insects are also captured when opportunity offers. Most of these animals, apart from large winged insects and young birds just learning to fly, are attacked on the ground. But when fresh prey is scarce, as may happen in a bad winter, the red kite has no hesitation in feeding on carrion.

Pairs of red kites return to the same breeding sites every year and even when one of the partners fails to appear the other seldom ventures far afield. The size of a small territory normally

RED KITE
(Milvus milvus)

Class: Aves
Order: Falconiformes
Family: Accipitridae
Length: 24½ inches (61 cm)
Wing-length: 19¼–20 inches (48–50 cm)
Wingspan: 56–64 inches (140–160 cm)
Weight: 1½–2½ lb (0·75–1·1 kg)
Diet: small mammals and birds, frogs, snakes, lizards, large insects; occasionally carrion
Number of eggs: 2–3 (exceptionally 1 or 4)
Incubation: 28–30 days

Adults
Chestnut tail deeply forked. Wings narrow, strongly angled, with large white patches on lower side. Back reddish-brown, head light brown, delicately striped with black. Cere and feet yellow.

Young
Fledgling covered with grey down on upper part of body, cream below. The growing bird is lighter brown than the adult, except for the head, which is darker. The tail is more deeply forked than that of the young black kite and its plumage contains more red than the latter.

varies from 500 yards or thereabouts to a couple of miles. But in districts where prey is abundant, the birds may roost so close to one another that they form small colonies which sometimes mingle with those of black kites.

First to arrive at the nesting site is the male, which perches on a branch and emits characteristic high mewing calls to attract his mate. Both birds later engage in a courtship display, consisting of a sequence of gliding flights high above the tree, in the course of which they exhibit a high measure of acrobatic skill. Now and then they interrupt the performance to couple, then resume the ritual, soaring into the sky and plummeting down towards the nest again and again.

The eyrie is usually situated in a tree, about 20–60 feet from ground level, but is seldom of the kites' own making. Very often it will have been built by another bird, such as a crow or buzzard, and is simply taken over by the breeding pair. If necessary they will collaborate in restoring the nest, the male bringing the materials, the female fitting them together. It may happen that when they are comfortably installed they will be driven out by a couple of black kites, which are in the normal way more active and aggressive.

In southern Europe the breeding season occurs in the first half of March and in central Europe roughly a month later. Two weeks after taking possession of the nest the female lays from one to four eggs (the average is two or three). These are whitish and sometimes speckled with brown or green. If there are several eggs, these are laid at intervals of about three days. Incubation lasts 28–30 days and when, as often happens, the female temporarily leaves her post, she is immediately replaced by her partner.

Of the two kite species, the red kite (*above*) is indisputably the more handsome, with its reddish-brown, elegantly striped plumage and its large amber eyes. The black kite (*below*) frequents watery habitats, but both species construct flimsy nests, lined with scraps of fabric, paper and débris, in the fork of a tree.

The black kite is smaller than the red kite and its tail is less deeply forked. Its plumage is also somewhat darker and more uniform in colour.

Facing page: The black kite is not an energetic bird, spending much time in the eyrie and then making good use of warm air currents for its characteristic gliding flight. The long wings are often bent and the forked tail is also a reliable guide to identification.

Because of the lapse of time between successive hatchings, the older fledglings are often considerably larger than the later arrivals. But being peaceably inclined at this age, the young birds do not fight with one another. Fatalities during this period are usually the result of starvation.

The fledglings grow fast and within a month are already covered with feathers, although the greyish down on the head persists for a while longer.

About six weeks after hatching the young kites are ready to leave the eyrie, having already learned to tear apart the prey which their parents continue to bring them, despite the fact that they are not yet fully grown. By now the birds are becoming more quarrelsome, and the stronger individuals do their best to appropriate the tastier food morsels.

When the youngsters are capable of flying they are accompanied and guided by their parents for about another month, during which time the family continue to occupy the eyrie.

The black kite

The black kite, far more numerous than its relative, is essentially a bird of watery habitats and is frequently seen describing broad circles in the cloudless sky above lakes and swamps. It is slightly smaller than the red kite and both sexes weigh between $1\frac{1}{2}$ and 2 lb. The plumage is basically dark brown rather than black, so that the common name is not strictly accurate. As the specific name suggests, this is a migratory bird, with winter quarters in tropical Africa. Ringed individuals have been retrieved in Sudan, Uganda, Togo, Ivory Coast, Ghana and Senegal. From the second week of August until the end of October most of these migrants fly south by way of the Straits of Gibraltar, although a few remain in Spain for the entire winter. The black kites inhabiting regions farther east cross the Bosphorus, while one small group passes over Sicily and Tunisia.

According to observations made by the eminent ornithologist Francisco Bernis, the first springtime arrivals from winter quarters occur throughout March, but they may begin prior to that in February or be delayed until April or mid-May. Certainly in central Spain the vanguard appears early in March. The kites glide in over cultivated fields and watercourses, only their forked tails making any perceptible movement. Within a fortnight they are occupying and defending their territories.

This is the time when they have their seasonal encounters with red kites, which are already incubating. Such meetings are not necessarily very violent, the adversaries usually lashing out with their feet and flying round each other, to the accompaniment of raucous cries. The black kites, which have evidently formed pairs prior to starting on their northward journey, often succeed, nevertheless, in dislodging the earlier arrivals.

As with so many other large birds, the true skill and beauty of the black kite can best be appreciated as it performs its aerial courtship display. Soaring to a great height, both partners describe great wheeling flight patterns; then one of them will hurtle downwards like a stone, almost brushing the ground

Black kites sometimes roost in colonies and each brood normally consists of two or three chicks. The fledgling pictured below is about one week old. Within a month it will have shed its down, grown feathers and begun flying.

before checking its fall and then resuming its climb, sometimes remaining poised in mid-air, absolutely motionless, in the manner of a kestrel.

The black kites begin to build their nest while they are still courting. This eyrie is generally situated on a branch or in a fork of an oak or pine, 15–60 feet from the ground and always close to water. Exceptionally it may be found on a rock ledge or building. It is a roughly constructed affair, little more than a shapeless mound of hastily assembled sticks and twigs, fortified with all manner of discarded odds and ends or waste matter (bits of cloth, pieces of paper, objects made of plastic, dried horse and cattle dung, etc).

The finished eyrie is so fragile and unstable that it is a wonder the tiny fledglings do not tumble out. The nest neither provides adequate shelter and concealment for the incubating female, nor a great deal of warmth and protection for the young birds. Indeed, the long tails of the fledglings can sometimes be seen from below, hanging over the edge of the nest.

The eggs (which vary in number from two to five, according to the region) are laid from the beginning of April to the end of May. They are creamy or whitish, adorned to a lesser or greater degree with reddish-brown spots of different sizes. The shells are smooth.

Much remains to be discovered about the reproductive habits of the black kite. It would appear, however, that the eggs are laid at two-day intervals and immediately incubated. The male does not show any sign of participating in this activity, catching food for his partner and standing guard on a nearby branch. The female occasionally abandons the clutch for a few minutes to venture on a short hunting expedition of her own.

The fledglings hatch 35–38 days after the eggs are laid. They weigh less than 2 oz but grow very quickly, feathers appearing within 16–18 days and then developing rapidly. By the time they are ready for their maiden flight, at the age of about five weeks, the youngsters weigh 1¾–2 lb. Although still fed by their parents, they gradually learn how to hunt for their own food which consists in the main of waste matter, carrion and moderate-sized animals, usually physically disabled individuals.

Black kites sometimes form small colonies, the nests being situated fairly close to one another. Although it may seem surprising that birds of prey of this size should be so gregarious, the tendency is not all that unusual when one bears in mind the raptor's astonishing powers of adaptability, with a diet that may range from insects, spiders and crustaceans to carrion. Nor does the bird recognise clearly defined territorial boundaries. One pair of kites will tolerate the presence of another couple but this does not mean that the birds necessarily live together harmoniously. There is a good deal of squabbling over food, and much pilfering of prey. No sooner does one kite discover a carcase than it is surrounded by a group of hungry companions.

When not fighting among themselves, the black kites are always ready to challenge other predators, harassing larger and stronger individuals and forcing them to abandon their victims. They boldly snatch fish from the bills of herons and systematically ransack the nests of these elegant wading birds.

A black kite on the hunt for food flies methodically over a stretch of open ground or water, this survey often taking it several miles from the eyrie during the breeding season. Expert at gliding, the bird will hover for hours in the air, scrutinising every square foot of ground until it sights a moving animal or the remains of another predator's meal. The prey is not necessarily large and the slightest activity on the part of a cricket, a grasshopper, a mouse or a rat will unleash a deadly, accurate power dive.

Although insects and all kinds of vertebrates are included among the black kite's victims, the predator has a particular partiality for eyed lizards and Montpellier snakes. When the water level in streams and rivers is low and fishes are stranded on the banks to die, the raptor is quick to take advantage of the offered delicacy; and although normally averse to wasting unnecessary energy, a more than usually agile individual will sometimes pounce on a rabbit affected by myxomatosis or even snatch a defenceless fledgling from its nest. Jesús Garzón Heydt suggests that rabbits and young birds will only be attacked by a pair of kites specialising in that form of hunting. This naturalist examined the contents of eleven black kites' nests and only in two did he discover the remains of several young herons.

Undoubtedly, the black kite is a useful bird of prey. It hunts insects and rodents that are notoriously harmful to crops, and performs a hygienic function by reason of its scavenging tendencies, reducing the risk of pollution and contamination by feeding on animal carcases and assuring the survival of many species by concentrating its attacks on weak and diseased individuals. The bird deserves to be legally protected.

Geographical distribution of the black kite.

BLACK KITE
(*Milvus migrans*)

Class: Aves
Order: Falconiformes
Family: Accipitridae
Length: 22–23 inches (56–58 cm)
Wing-length: 16½–18½ inches (41·5–46·5 cm)
Wingspan: 44–58 inches (110–145 cm)
Weight: 1½–2 lb (660–927 g)
Diet: very varied—insects, small birds, fishes, sometimes carrion
Number of eggs: 2–4, rarely 1
Incubation: 35–38 days

Adults
Similar to red kite but tail less prominently forked. Upper parts dark brown. Chest and belly reddish-brown with black marks. Eye chestnut-yellow. Cere and feet yellow.

Young
Fledgling covered with whitish down. Later there are small white marks at the base of the primary remiges.

CHAPTER 3

The rodents and their foes

The Mediterranean region is not renowned for a concentration of small mammals, although it is possible that before the outbreak of the myxomatosis epidemic there were as many rabbits in the area as there are prairie dogs on the Great Plains or susliks and marmots on the Eurasian steppes. Yet even though the maquis is not distinctive for the size and variety of its small mammal population, it is nevertheless characterised by an extremely interesting and highly adaptable mixed community of rats, mice, dormice, fieldmice and similar rodents.

Although the disaster which overtook the rabbit population initially benefited the various rodent species in that it removed their principal rivals for food, the long-term effect was very different. Predators that had previously relied mainly on rabbits for sustenance now turned their attentions to the teeming world of rodents. Foxes, wildcats and weasels – all woodland hunters – complemented the activities of numerous nocturnal birds of prey. In addition to these predators are two carnivores in parts of southern and south-western Europe, the ichneumon or Egyptian mongoose and the genet – the only local representatives in Europe of the Viverridae. Both are more properly African and the two differ from each other not only in colour and general appearance but also in hunting techniques, the former capturing prey on the ground, even in regions of dense vegetation, the latter usually seeking its victims in trees.

Of all the rodents of the Mediterranean, the most conspicuous and beautiful is without doubt the garden dormouse (*Eliomys quercinus*) – misleadingly named because far from frequenting gardens it is a typical inhabitant of thickly vegetated woodland. Although found throughout continental Europe, this engaging little animal is particularly abundant in these southern climes.

Facing page : The garden dormouse is the most typical rodent of the woods and forests of the Mediterranean region. It generally spends the day in a hollow branch and ventures out at night to hunt insects, small mammals and birds or to lay in a stock of acorns and chestnuts for the winter.

The garden dormouse is distinguished for its strikingly contrasted colours and tufted tail. The latter may come to its aid against an enemy, for the fragile skin near the tip breaks off in the predator's mouth, the tail proper being discarded a little later. It is rare, however, for a new tail to develop.

GARDEN DORMOUSE
(Eliomys quercinus)

Class: Mammalia
Order: Rodentia
Family: Gliridae
Length of head and body: $4\frac{1}{2}$–$7\frac{1}{4}$ inches (11–18 cm)
Length of tail: $3\frac{1}{2}$–$4\frac{3}{4}$ inches (9–12 cm)
Length of foot: $\frac{3}{4}$–$1\frac{1}{4}$ inches (2·2–3·2 cm)
Weight: $1\frac{1}{2}$–5 oz (45–140 g)
Diet: mainly fruit, seeds and other vegetable matter; also insects and other small animals
Number of young: probably 2 litters annually, each with 2–7 babies

Adults
Colour of back variable, ranging from dark grey to reddish-brown. Belly paler. A black patch on the head, starting behind the muzzle, encircles the eyes and continues under and beyond the ears, which are comparatively large. The long tail terminates in a tuft of hair; usually it is black above and white below.

Young
Naked and blind at birth, eyes opening after about three weeks.

Its coat colour is a striking contrast of russet or grey above and white below; and a jet-black patch frames and blurs the outline of the eyes, extending behind the large ears.

The most extraordinary feature of this handsome animal, however, is its long tail, ending in a tuft, the colour of which varies according to race (in the southern population, particularly in Spain, the lower side has black markings, but in animals living farther north the underside is white). The dormouse's tail has several functions. It helps of course to balance the body, steadying it for running, jumping or climbing; it serves as natural camouflage, for the dormouse can simply curl it around the body when it is asleep so that natural outlines and angles are obscured; and–most astonishingly–it can act as a snare or booby-trap. If surprised by an enemy, the dormouse reacts by waving the tail violently in all directions, deliberately attracting the predator's attention. When the latter attempts to seize its victim, the last few inches of tail suddenly snap off so that the tip is left between the attacker's teeth, giving the dormouse time to make its escape.

This phenomenon is also common to certain species of lizards, but here the process does not seem to be quite the same. The dormouse's tail does not break off completely and only the outer skin is ruptured at this stage, the dried-out and unprotected tip dropping off of its own accord a few days later. The mutilated tail may grow again but apparently this is a fairly rare occurrence.

The garden dormouse, although predominantly vegetarian (seeds and fruit), is in fact an omnivore. Thus in its hunting capacity it consumes numerous insects, snails and smaller mammals, as well as eggs and fledglings snatched from the nest.

Scientists concerned with the International Biological Programme (IBP) have considered the possibility of trying to introduce the garden dormouse into pine woods because of its partiality for processionary caterpillars, which do such extensive damage to these conifers.

The garden dormouse is of course itself hunted by a horde of carnivores, including wild cats, foxes, genets, ichneumons, barn owls, tawny owls and the long-eared and short-eared owls. Should it be cornered, the rodent tries to use its teeth in self-defence (the incisors are quite powerful and capable of inflicting painful bites), but it more often manages to escape by running away as fast as it can. It is an accomplished climber but generally spends most of its time at ground level, sleeping between stones or in hollow tree trunks. Sometimes it makes use of an abandoned nest–usually that of a magpie–providing it with a fresh layer of moss, working it into a circular shape and excavating a new side entrance. It is in such a refuge that the dormouse–either alone or in a group–settles down for its winter sleep. This is a dangerous season for other species (the local fat or edible dormouse population, for example, may be reduced by four-fifths during the winter) but the garden dormouse offsets this winter mortality by a very high rate of reproduction. Each female may have two litters a year, both of which comprise up to seven babies. In fact the precise number of litters is not known and appears to be affected by several factors which vary from one region to another.

The dreaded plague carriers

Several thousand years ago the brown rat or Norway rat (*Rattus norvegicus*) was only found in the deserts and steppes of central Asia. But it was not long before this prolific species extended its geographical range. Aelian mentioned invasions of rats north of the Caspian Sea in the 3rd century. Gessner described the species in 1550, which suggests that they had reached Europe long before Pallas saw them in 1727 swimming across the Volga in their thousands, following an earthquake. According to Hainard the brown rat was first introduced into England in 1730 in the holds of ships from Norway. It was discovered in Paris in 1793 and in Switzerland in 1809. After this it rapidly spread all over the European continent. Today it has a worldwide distribution.

How is it that these rodents succeeded in establishing themselves in so many different habitats? Undoubtedly their lack of specialisation was invaluable in permitting them to survive in any situation and conditions, but an equally important factor was their pattern of social organisation, which modern science has shown to be grimly efficient. The German naturalist F. Steiniger has conducted interesting experiments with brown rats and has described how he placed a number of subjects–unrelated to one another–in a fairly extensive enclosed area. At first the animals, obviously frightened, wandered at random around the park, studiously avoiding one another, although those that happened to meet on the outskirts clashed violently. In due course the large males began staking out frontiers and sallying forth in search of mates. Pairs of rats were soon challenging other couples and mercilessly attacking solitary individuals. In the course of long, bloodthirsty battles the stronger rats pitilessly exterminated the

The house mouse which originated in eastern Asia has since spread all over the world. In many countries it has become a serious pest since it feeds on seeds, grains and many kinds of crops. The more its natural predators, both on the ground and in the air, are decimated, the more likely the species is to assume plague proportions in the near future. The only sensible way of keeping the population under control is to preserve and encourage the proliferation of these natural enemies.

Facing page: Like all rodents, black rats (*above*) are extremely prolific, capable of having up to six litters a year, each of which may comprise ten young. They infest both town and country and have a worldwide distribution. Wood mice (*below*) are, as their name suggests, primarily rodents of wood and forest but are also found in gardens and cultivated land; in winter they often seek refuge in barns.

weaker ones. In Steiniger's words, 'They would move forward slowly and furtively, displaying extreme caution, and then pounce on the unsuspecting animal, usually as it was feeding, aiming at either side of the neck to sever the carotid artery. Such encounters normally lasted only a few minutes, as the mortally wounded victim bled to death'.

In the course of these deadly assaults each dominant rat seems to pick out an animal of its own sex. The sequence of fights continues until only one pair of rats is left alive. These proceed to reproduce so that they alone are responsible for fostering a huge rodent colony. The curious thing is that the bitter rivalry pattern is not perpetuated in subsequent generations. One might assume that once the offspring have attained adult status they will in their turn be tormented and persecuted by their parents, but this is not the case. Hereafter the rats seem to live together in perfect harmony, and the original victorious dominant will even share his partner with other males.

Judging by this spectacle of tribal loyalty and amity, it would seem logical to suppose that the elders of the community, having once satisfied their lust for blood, might extend this tolerance to strangers. This, however, is dramatically disproved by introducing an alien rat from another family into the group, when the carnage begins again. The question which then arises is whether the different members of the same family actually recognise one another, and if so how this is achieved. Steiniger isolated one rat from its companions and later reintroduced it to the colony. It was immediately attacked and killed, the reason being that this individual had lost its 'collective odour', proving that members of such a community know one another by smell and that each rat has its own identification sign.

Can a selective mechanism which only allows one reproductive pair of a species to survive really be considered efficient? The size of the population provides the answer, for of course any adverse repercussions of this intraspecific warfare are more than offset by the extraordinary fertility of the survivors. One pair of rats, between spring and winter, is capable of producing seven litters, each comprising five to twenty babies. The latter are born red, naked and blind, and are less than two inches long. Within about two weeks their eyes open and when they are three months old they are themselves capable of reproducing.

Although brown rats generally live in towns—at man's expense—they are also found in the country on the banks of streams and rivers, for they are semi-aquatic. In the Cota Doñana reserve, where they play as important a role in the food chain as do the garden dormice, they consume large numbers of frogs and toads, usually destroying their eggs and devouring the young. But their habit of only eating a part of their prey and leaving them dead or badly injured benefits the local birds of prey. Dr Valverde described how in July 1956, a year in which toads were particularly numerous, he would find the edges of the swamp littered every morning with dead or dying amphibians, most of them horribly mutilated, some without hind legs, these having been wrenched away, others with dreadful wounds on chest and hindquarters, the viscera having been removed. A few were

RATS

Class: Mammalia
Order: Rodentia
Family: Muridae

COMMON, BROWN OR NORWAY RAT
(*Rattus norvegicus*)

Length of head and body: $8\frac{3}{4}$–$10\frac{1}{4}$ inches (22–26 cm)
Length of tail: $6\frac{1}{2}$–8 inches (16·5–20 cm)
Weight: 10 oz–$1\frac{1}{4}$ lb (275–580 g)
Diet: practically omnivorous
Number of young: 2–7 litters annually, each with 5–20 babies

Upper parts greyish-brown, lower parts greyish-white. Blunt muzzle; ears hairier but less well developed than in black species; tail shorter than rest of body and composed of 160–190 rings.

SHIP OR BLACK RAT
(*Rattus rattus*)

Length of head and body: $6\frac{1}{4}$–$8\frac{3}{4}$ inches (16–22 cm)
Length of tail: $6\frac{3}{4}$–$10\frac{1}{4}$ inches (17–26 cm)
Weight: 5–9 oz (145–250 g)
Diet: practically omnivorous
Number of young: 3–6 litters annually, each with 4–10 babies

Adults
Smaller and more slenderly built than brown rat; pointed muzzle, large ears, tail generally longer than rest of body and composed of 200–260 rings. Colour dark, in various shades of grey.

Young
Young of both species are born naked and bright red. They grow rapidly and are capable of reproducing at three months.

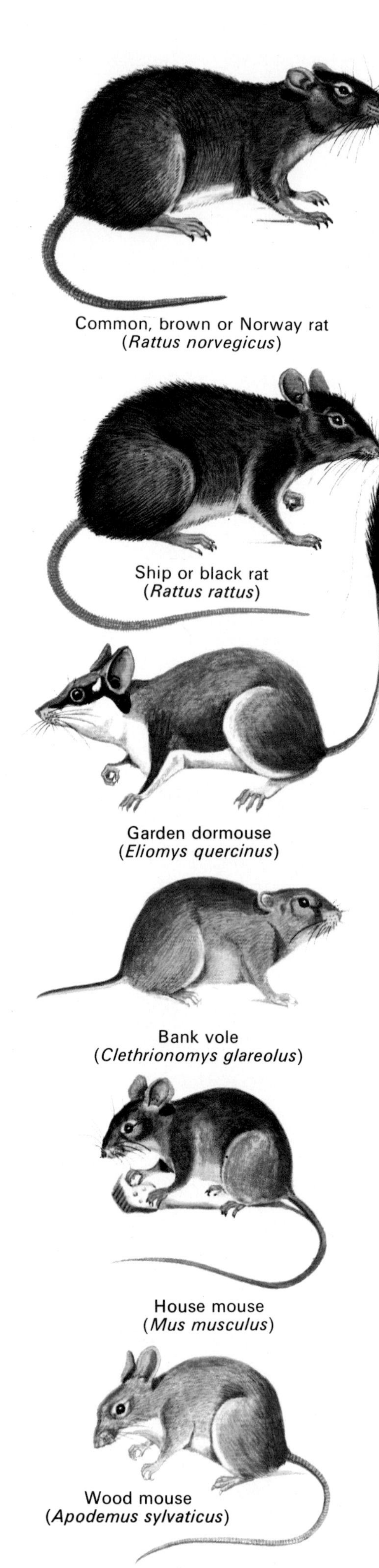
Common, brown or Norway rat
(*Rattus norvegicus*)

Ship or black rat
(*Rattus rattus*)

Garden dormouse
(*Eliomys quercinus*)

Bank vole
(*Clethrionomys glareolus*)

House mouse
(*Mus musculus*)

Wood mouse
(*Apodemus sylvaticus*)

still able to drag themselves painfully about in the grass, eventually being scooped up by low-flying black kites. By mid-day they had all been devoured either by raptors or ants.

The closely related black rat (*Rattus rattus*) also originated in Asia and had made its appearance in Europe, including Britain, by the Middle Ages. But reports of the discovery of its bones in prehistoric sites would seem to indicate that it might have been a much earlier arrival on the Continent. Its habits are very similar to those of the brown rat, which appears to have supplanted it in some regions. Research suggests that each species assumes temporary dominance in a given environment, for reasons which are not so far explained. In parts of Switzerland, for example, the black rat was ejected by its brown relative, only to make a subsequent reappearance. The black rat is in any event more of a country dweller than the brown species, often establishing a lair high up in a pine. It is also less carnivorous.

Both the brown rat and the black rat, however, have plagued mankind, figuratively and literally, throughout its history. As carrying agents of bubonic disease, typhus and rabies, these rodents have been responsible for the world's most terrible epidemics, including the Black Death. Fortunately, modern science and hygiene have at last found the antidotes to rat-borne pestilence. Nevertheless, rats continue to cause widespread harm and misery both in town and country, and all attempts to exterminate them have failed.

Related to the rats, also destructive to crops but less dangerous to health, is the house mouse (*Mus musculus*), likewise an inhabitant of urban and rural areas. According to Irenäus Eibl-Eibesfeldt, these rodents are extremely aggressive towards other mice of different species and engage in violent territorial battles. The same scientist points out that each animal has a separate nest and several lairs. There is a rigid hierarchy among the males and when they fight they rear upright on their hind legs and lash out viciously with their front legs.

The Iberian root vole (*Pitymis duodecimcostatus*) is another member of the family which spends at least half its time underground and poses a serious menace to cultivators of orange and lemon groves. But by far the most common rodent of the Mediterranean region is the wood mouse (*Apodemus sylvaticus*) which is also, according to G. S. Miller, the most numerous mammal in Europe. This highly prolific rodent is, not surprisingly, preyed upon by every kind of carnivore and raptor.

The striped and spotted hunter

There was probably a time, long ago, when the world's rodent population was maintained at a tolerable level, thanks to the activities of carnivores alone. If so, man's arrival on the scene soon changed this delicate balance and from then on he was doomed to fight a losing battle against an ever-increasing army of rodents. He started by domesticating certain traditional predators and enlisting their aid in the uneven contest. Thus the cat was initially tamed in Egypt and thereafter worshipped as a sacred animal. But it was not until the 9th century AD that it found its

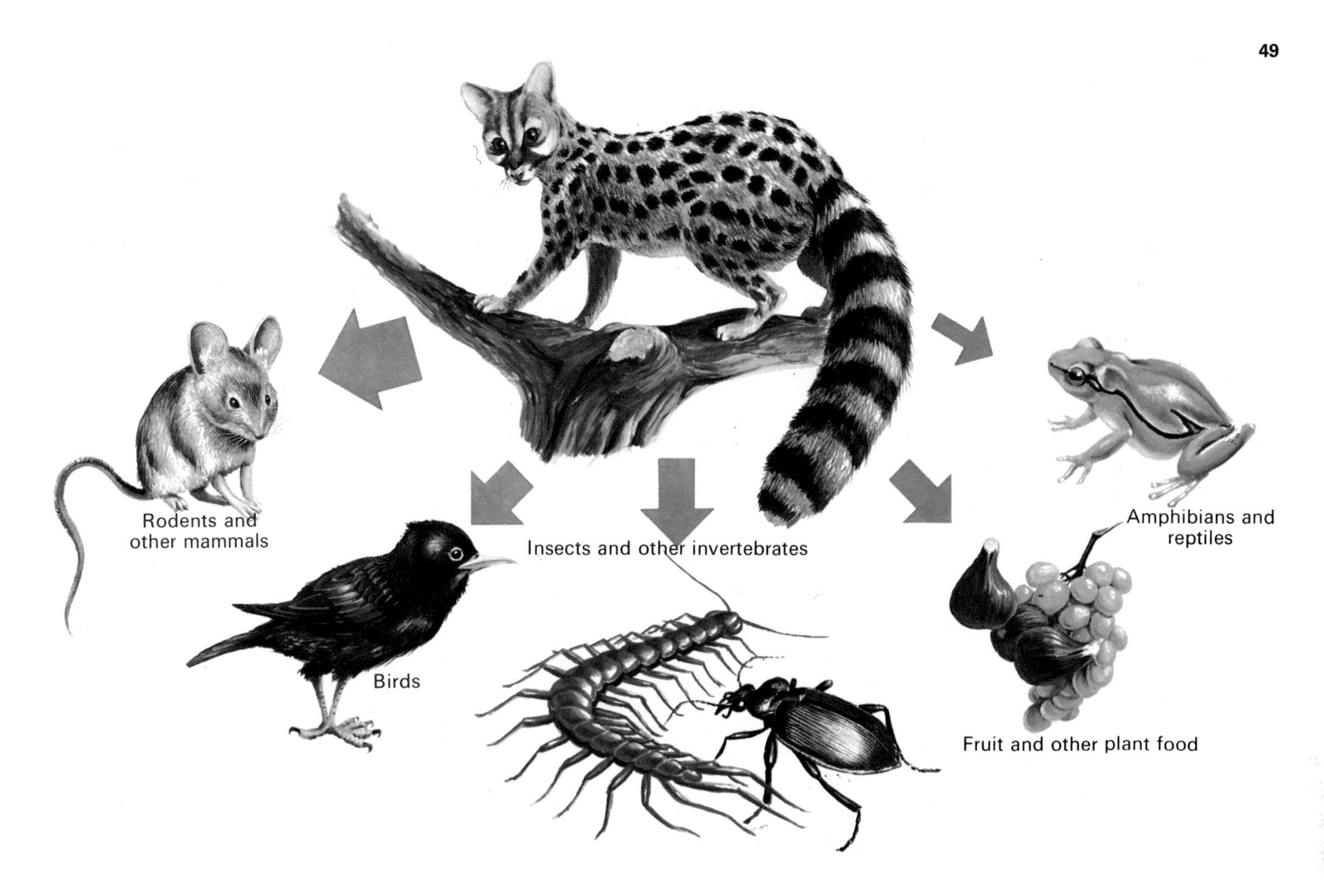

way into western Europe. Prior to that it appears that many households plagued by rodents made use of two animals nowadays regarded as harmful species–the weasel and the genet.

In the opinion of a number of authors, the small-spotted genet (*Genetta genetta*), as slender and beautiful as a cat, can be quite easily domesticated. One important reason why it was superseded as a household ally and pet by the cat was doubtless because it gave out an unpleasant, though harmless, musky odour, exuded from the glands on either side of the anus.

Nowadays the genet is seldom seen in the country, largely as a result of its nocturnal habits, for it comes out to hunt at dusk and retires to its lair at dawn. But its forest origins are clearly revealed by the distinctive coat markings–a splendid mixture of black spots, with black rings on the tail, against a pale ground–an ideal protective pattern for a hunter which slips silently along the branches of an oak, through foliage dappled by the moonlight. The genet's large, dark eyes are also perfectly suited for nocturnal hunting expeditions. Wherever it prowls–in woods, shrubs or undergrowth–this agile carnivore can depend on almost total concealment from its prey.

Sometimes the graceful genet slinks furtively between rocks or through low-lying foliage, watching, waiting and then suddenly pouncing on anything that moves–the very model of an opportunistic predator with a broad, unspecialised food range. But in certain situations it will utilise very different methods, for it is nothing if not adaptable.

Evidence of the animal's versatility has been shown by placing a genet in a large enclosure, room or cage with which it is unfamiliar. This confined space is deliberately strewn with branches and other natural obstacles, and all these will immediately be

Food preferences of the small-spotted genet. As the chart shows, this European member of the Viverridae feeds mainly on rodents, next in order being birds, insects and other invertebrates. It is a nocturnal hunter, with excellent vision and cryptic coloration.

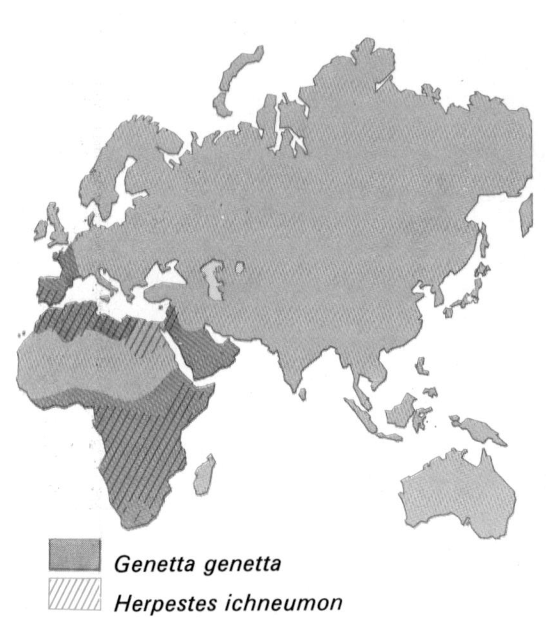

Geographical distribution of small-spotted genet (*Genetta genetta*) and ichneumon or Egyptian mongoose (*Herpestes ichneumon*).

SMALL-SPOTTED GENET
(*Genetta genetta*)

Class: Mammalia
Order: Carnivora
Family: Viverridae
Length of head and body: 18$\frac{3}{4}$–24$\frac{3}{4}$ inches (47–62 cm)
Length of tail: 15$\frac{1}{4}$–19$\frac{1}{4}$ inches (38–48 cm)
Weight: 2$\frac{1}{4}$–4$\frac{3}{4}$ lb (1–2·2 kg)
Diet: rodents, birds, insects and other invertebrates, fruit etc
Gestation: 10–11 weeks
Number of young: 1–4, usually 2

Adults
Long body with short legs. Long, hairy tail, ringed black-and-white. Colour light brown or greyish with black spots and patches. Black facial mask with small white marks. Completely black individuals not uncommon.

Young
Darker coat colour, already spotted at birth.

Facing page: The various members of the genus *Genetta* are African Viverridae, but the small-spotted genet also is found in south-western Europe, with some isolated individuals in the German forests. Although they have longer bodies and hairier tails they resemble slender, delicately built cats in appearance and their hunting methods are similar. Furthermore they are very sociable and can easily be domesticated.

subjected to the most careful scrutiny. With extreme caution and precision, the genet measures its every step, refraining from bringing the entire weight of its body to bear on any object until it is absolutely certain of its stability. In this way it familiarises itself with every branch and stone which goes to make up its environment. Having made a complete circuit of its allotted territory, sniffing at every corner, the genet will then begin another exploratory journey, though this time displaying less hesitation and thoroughness.

After it has trotted round two or three times, always by the same route, the animal begins to move about surprisingly fast, jumping from one rock or branch to another, almost seeming to rebound from the surface, merely grazing it. By now every movement reveals its complete assurance.

As a result of such experiments scientists are satisfied that eyes, whiskers, ears and nose all play an important part in territorial exploration and location. Firstly there is absolute visual recall of the terrain, for the eyes are capable of picking out immobile objects in almost complete darkness. Then there is tactile familiarity, with the aid of the vibrissae on the muzzle. The large ears help the animal to detect the faintest sounds; and the sense of smell, though not as good as that of the dog family, is sufficiently well developed for location purposes, and apparently much better than that of other members of the family Viverridae and of most felines.

Maurice Burton points out that the genet's attachment to certain fixed routes is so marked that in the event of its having tumbled from a branch on its first attempt and being forced to make a détour underneath, it will studiously avoid this particular obstacle in all its subsequent travels, even if it is kept in the same familiar cage for a number of years. There is good reason to believe that it behaves similarly in the wild.

Rodents make up by far the greatest proportion of the genet's diet. In fact, out of 28 stomach contents' analyses, 19 revealed remains of rats and mice (one contained four rodents). The victims, however, are by no means defenceless, for experiments have proved that rodents also move about rapidly and confidently in semi-darkness and that they too retain an astonishing visual memory of the details of their surroundings. One such test featured a mouse feeding outside its burrow during the early hours of the night. Naturalists had carefully removed all the branches, twigs and stones which normally formed obstacles in the path of the rodent as it fled, pursued by an enemy, towards its lair. Having cleared the ground, films were taken by infra-red light of the mouse which, since it could not see the camera, did not vary its usual behaviour. As it moved backwards and forwards across the area, the animal executed little leaps and sideways steps to avoid the imaginary obstacles, exactly as if they were still in their customary positions.

The hunting technique of the small-spotted genet more closely resembles that of a cat than that of the other local representative of the Viverridae–the ichneumon. Normally it stalks its prey, belly flattened against the ground, then pounces unexpectedly on its victim. But it has such a varied choice of food that it is

A family of ichneumons will often march in single file, head under the tail of the animal in front.

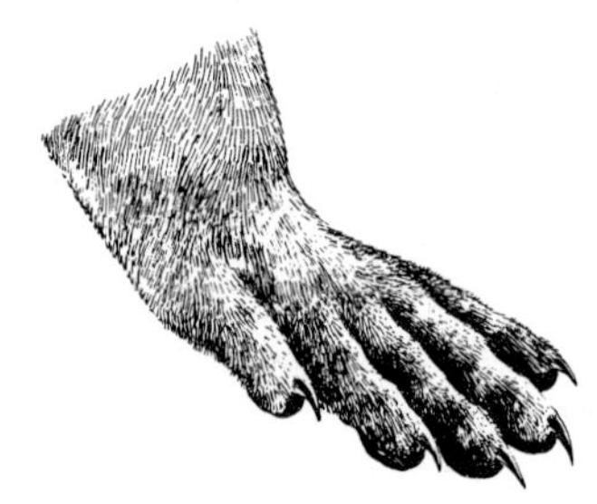

The toes of the ichneumon are equipped with non-retractile claws. They are less powerful than those of the cat family but longer and sharper than those of dogs and thus useful both for offensive and defensive action.

ICHNEUMON OR EGYPTIAN MONGOOSE
(*Herpestes ichneumon*)

Class: Mammalia
Order: Carnivora
Family: Viverridae
Length of head and body: $20\frac{1}{2}$–22 inches (51–55 cm)
Length of tail: $13\frac{1}{4}$–18 inches (33–45 cm)
Diet: rodents, rabbits, reptiles
Gestation: 11 weeks
Number of young: 1–4

Adults
Streamlined body with short legs. Greyish-yellow pelage but little hair on lower parts. Colour of back darker. Tail thick at root, tapering to tip and terminating in tuft of black hairs.

Young
Very similar to adults.

virtually impossible to determine a preference for any one form of prey in a particular district, or at any special season. It is true that when fruit is ripe the genet will raid pear and fig trees and that in winter it seems to concentrate on birds and mice. Country-dwellers have remarked on the tendency of a genet to return to the same hunting ground every season, but this has not been scientifically verified. In the course of one night, a single animal may enjoy a mixed meal that includes pears, mice, a bird, a shrew and a couple of scorpions, indicating pretty clearly that it will feed on more or less anything that happens to be available.

The genet usually hides in the cleft of a rock or (more frequently) in the hollow of a tree. Here the female gives birth to her litter, generally comprising two or three (sometimes one or four) babies. Intruders are kept at bay as the carnivore erects its hairs, emits menacing growls and if necessary bares its fangs. Thanks to information provided by Dr J. Volf in the Prague Zoo, it is known that gestation lasts ten or eleven weeks and that births usually take place during the spring, although they may occur at any time of the year.

The ichneumon or Egyptian mongoose

The ichneumon (*Herpestes ichneumon*) is, like all other members of the mongoose family, a compactly built and rather elegant animal. The alternative name – Egyptian mongoose – derives from the fact that it was once domesticated in Egypt and regarded as a sacred animal. Apart from the small-spotted genet, it is the only European representative of the Viverridae and like the former it is an unspecialised carnivore and predator. It moves slowly on short legs, seldom halting its advance but carefully examining every bush, stone, hole or heap of wood in its path. As soon as it sights a likely prey it comes up from behind to launch an attack from short range, but if the victim tries to escape the mongoose will not usually trouble to chase it. Collective hunting is not uncommon, since family feeling is strong. Dr Valverde reports that in the Cota Doñana reserve recent surveys show that about half the ichneumon population hunt on their own, the remainder venturing out in pairs or in groups of from three to seven individuals.

When hunting in bands, the mongooses cover every square yard of ground, continuously uttering little cries so that they all keep close together. It is very common to see one animal at the head of a column, each of its companions following immediately behind the animal in front, head tucked under the latter's tail. Swaying slightly on their stumpy legs, which are half-hidden under the long fur, the animals resemble a huge caterpillar. A hunter from Andújar in the Sierra Morena – one of the last regions

where the animals are still found in any number—told with amazement how he once came across a 'dragon biting its tail', apparently turning around in endless circles. He had evidently surprised five baby ichneumons, the mother having fled at his approach and left her terrified offspring to take cover in the traditional manner, each hiding its head under a sibling's tail. They thus formed a complete circle which none dared to break, at the same time running forward blindly as fast as their legs could carry them. Then as the hunter came nearer to get a better view of the extraordinary sight, the little animals suddenly broke loose and dashed off—in the observer's words, 'like a snake that had cut itself into pieces, each with a separate existence'.

Ichneumons live in the densest areas of the Mediterranean maquis and do not venture near pine and oak woods unless there is a sufficiently thick cover of underbrush. In the Doñana reserve they appear to favour districts planted with lentisk, but they are also found among cistus, brambles, rushes and reeds that border marshes and swamps.

Although it was long assumed that Egyptian mongooses were nocturnal animals and that they fed mainly on rabbits, Dr Valverde's research colleagues have corrected these notions, especially the former. The little mammals begin to busy themselves at dawn, and although it is no easy matter to spot them among the dense vegetation, they are apparently out and about at intervals until nightfall. They are most active between 10 am and 2 pm, but less so during the next three hours—normally the hottest part of the day. A few animals (five out of a total of 178) were sighted between 9 pm and 10 pm but the majority had long since disappeared and there was no sign of them until about 6 am the following morning. It is true that location of the animals was more of a problem in the darkness and that the scientists spent more time working in the daytime, but there is no disputing the conclusion that these mammals are basically diurnal by habit and that they always hunt when it is light.

The Doñana scientists were also able to prove that the mongooses' food range is far more eclectic than was previously believed. Although rabbits certainly feature prominently in their diet (they often catch them by imitating the alarm calls of the latter when wounded), hunting grounds do not noticeably co-

Although long considered a nocturnal carnivore, the ichneumon is now known to go hunting by day as well. It glides silently through the grass in search of rodents, never pausing until it hears its prey moving. It then rears up on its hind legs and surveys the surroundings.

Among the largest reptiles of the Mediterranean region are the ladder colubrid (*above*) and Montpellier snake (*below*). Both are formidable predators with a wide food range but are not dangerous to man.

incide with those areas where rabbits are most abundant. It is confirmed that, like their Asiatic relatives, the ichneumons of the Mediterranean capture reptiles, including lizards and mice. A detailed survey would doubtless reveal an even more broadly based hunting pattern. G. A. Seaman and J. E. Randall have shown that a closely related species, the small Indian mongoose (*Herpestes auropunctatus*), which has been introduced into the Antilles, preys on fawns of one of the mule deer of genus *Odocoileus*.

Little is known of the ichneumon's reproductive behaviour, because the animal is seldom seen actually mating (either in April or May) or giving birth, after a gestation of eleven weeks.

In Europe the Egyptian mongoose is chiefly found in the south-western part of the Iberian peninsula and in Dalmatia, the assumption being that it was introduced into both regions.

Mediterranean snakes and lizards

No list of predators of the Mediterranean region would be complete without mentioning snakes and lizards, whose victims range from tiny insects to rabbits and animals of similar dimensions. These reptiles are in turn hunted by mongooses and other carnivores as well as by diurnal and nocturnal birds of prey, notably the short-toed eagle, which feeds on little else. Another interesting point about the snakes of the region is that they often attack others of the same species, particularly smaller individuals.

The Montpellier snake (*Malpolon monspessulanus*) is probably the most striking example of an animal which plays a shifting role in the regional ecology, for it is by turns prey, predator and superpredator. This is the largest snake in southern Europe, found in Spain, Portugal, southern France and parts of Italy, as well as in North Africa. The adults are olive-green with a distinctive black line down the spine. Some individuals may grow up to 8 feet in length. A separate race is found eastwards as far as Iran and retains throughout life the multicoloured pattern shown only in the young of the typical species.

The Montpellier snake is an active and frequently aggressive reptile. Coupling generally takes place in spring and eggs are laid at the beginning of summer. Each female apparently lays 8–12 eggs, which are approximately 2 inches long.

Shortly after they are born the young snakes start to hunt all kinds of insects and soon progress to small lizards, two types of prey which constitute their principal food during the first few years. Dr Valverde reports that between the ages of five to seven years the snakes regularly feed on mice and amphibians as well, and that when they are about ten years old–by which time they measure 3–4 feet–they also capture fledglings of ground-nesting species. From that time onwards–and particularly when they approach their maximum size at the age of eighteen years or thereabouts–they will attack other snakes of the same species. Dr Valverde came across a nineteen-year-old Montpellier snake that was not especially large (a little over 4 feet long), yet which had swallowed a seven-year-old individual almost 3 feet long.

Among other animals regularly selected as prey are eyed

MONTPELLIER SNAKE
(*Malpolon monspessulanus*)

Class: Reptilia
Order: Squamata
Family: Colubridae
Length: up to 100 inches (250 cm)
Diet: reptiles, insects, mammals, birds
Number of eggs: about 10

The typical subspecies is olive-green with a black mark down the back which becomes clearer as the reptile grows. The young display a multicoloured pattern similar to that shown by both young and adults of the eastern subspecies.

LADDER SNAKE
(*Elaphe scalaris*)

Class: Reptilia
Order: Squamata
Family: Colubridae
Length: up to 64 inches (160 cm)
Diet: small mammals, insects, birds, reptiles
Number of eggs: 8–10

Back grey or yellowish with two darker longitudinal lines. In the young, and sometimes the adults too, these lines are linked by thin transverse stripes, giving a ladder-like effect.

LATASTE'S VIPER
(*Vipera latastei*)

Class: Reptilia
Order: Squamata
Family: Viperidae
Length: up to 24 inches (60 cm)
Diet: lizards, small mammals etc
Number of young: 8–12, born live

Small horn on snout, facing slightly backwards. Greyish or reddish with a pattern of dark lozenges and marks on either side of the back. Belly varies in colour from yellow with grey spots to dark grey.

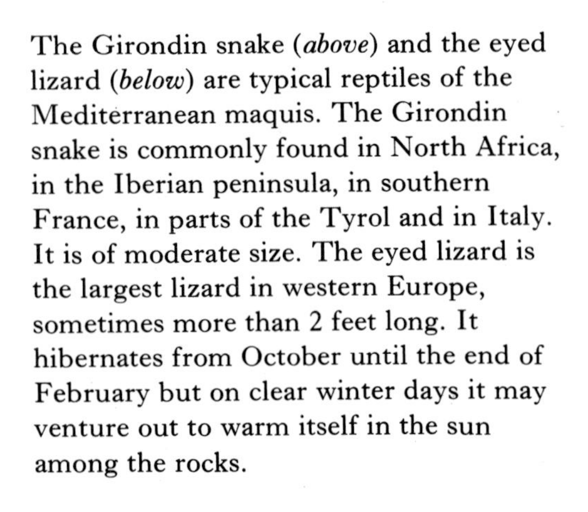

The Girondin snake (*above*) and the eyed lizard (*below*) are typical reptiles of the Mediterranean maquis. The Girondin snake is commonly found in North Africa, in the Iberian peninsula, in southern France, in parts of the Tyrol and in Italy. It is of moderate size. The eyed lizard is the largest lizard in western Europe, sometimes more than 2 feet long. It hibernates from October until the end of February but on clear winter days it may venture out to warm itself in the sun among the rocks.

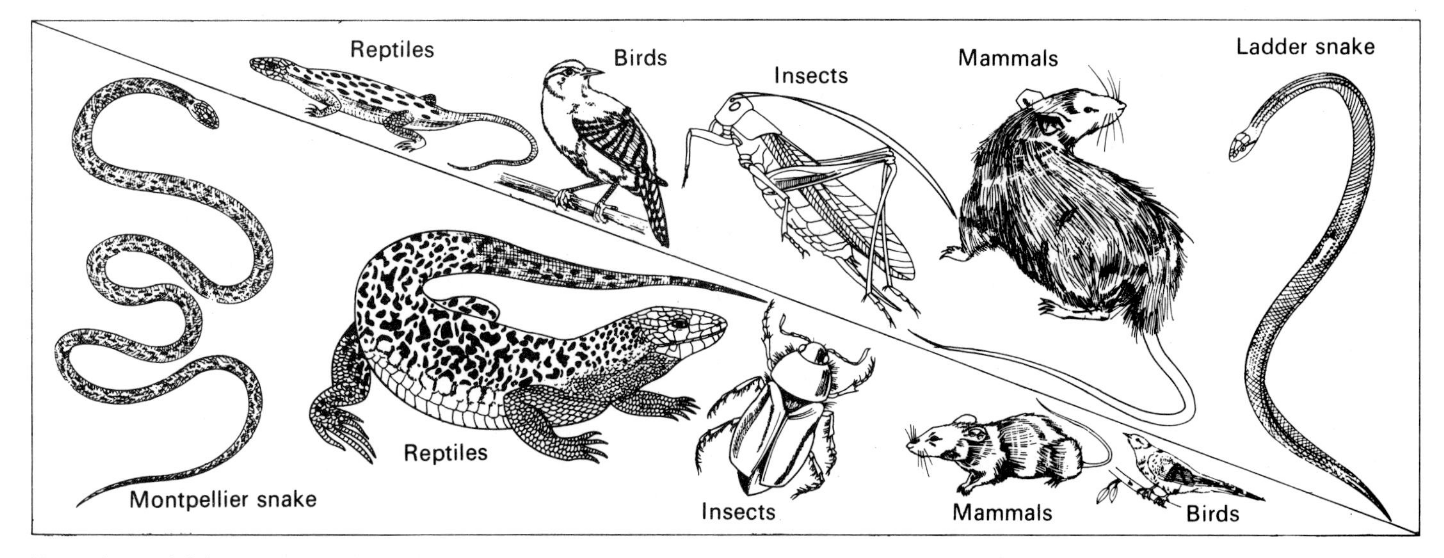

Although the ladder snake and the Montpellier snake hunt the same kinds of animals, they do not usually enter into direct competition with each other for the former preys principally on rodents and the latter on other reptiles.

lizards, rabbits, young herons and many other species of birds as well as birds' eggs. Bee-eaters are particular favourites, the snakes wriggling into the tunnels which the former drill into sandy embankments during the breeding season. But although the Montpellier snake kills many animals of ecological value and interest, it performs a service by helping to control the much more destructive rat and mouse population.

Another snake which is an equally determined hunter of ground vertebrates is the ladder snake (*Elaphe scalaris*), so named because of the two longitudinal lines down the back which are linked by transverse bands–the ladder-like effect being especially marked in younger individuals. This snake is found only on the Mediterranean shores of France, in the Iberian peninsula and in the Balearics. During its earliest years it feeds principally on insects but by the time it is three or four years old it has graduated to an exclusive diet of warm-blooded vertebrates, nine-tenths of them rodents.

Other species of the genus *Elaphe* are to be found along the Italian coasts, in Greece and in other parts of the Balkans. They include the four-striped snake (*Elaphe quatuorlineata*) and the leopard snake (*Elaphe situla*). The Aesculapian snake (*Elaphe longissima*) is a woodland species with a particular liking for deciduous trees; and Lataste's viper (*Vipera latastei*) and the Girondin snake (*Coronella girondica*) are characteristic reptiles of the warmer countries of Europe, notably southern France, as well as North Africa.

The eyed lizard (*Lacerta lepida*) is one of the most important reptiles of the Meditarranean area and in fact the largest lizard in western Europe, often measuring more than 2 feet long. It feeds on insects, fruit, molluscs and (in some experts' opinion) on small vertebrates. The reptile is in turn hunted by short-toed eagles, booted eagles, kites and buzzards.

Among other typical inhabitants of the thickets and sandy expanses of the shoreline bordering the Mediterranean are the common gecko (*Tarentola mauritanica*), several land tortoises, such as the Iberian or Algerian tortoise (*Testudo graeca*), and a variety of wall lizards (*Lacerta hispanica, Psammodromus algirus* and *Acanthodactylus erythrurus*).

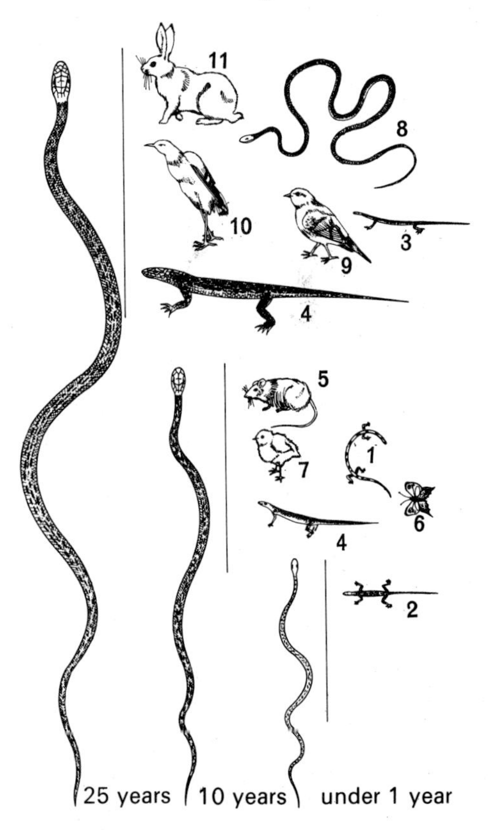

This diagram, based on the findings of Dr Valverde, shows the animals which appear regularly in the diet of the Montpellier snake. They include the skink (1), Spanish lizard (2), fringe-toed lizard (3), eyed lizard (4), rodents (5), larvae and adult insects (6), various young birds (7), snakes (8), medium-sized and large birds (9–10), and rabbits (11).

FAMILY: Viverridae

The Viverridae–animals which in their appearance bear a close resemblance to the very primitive and now-vanished group known as Miacidae–are carnivores belonging to the suborder Fissipeda (terrestrial hunters with paws and toes, in contrast to the aquatic Pinnipedia with paddle-like feet) and to the superfamily Feloidea–mammals distinguished by a partitioned tympanic (ear-drum) cavity.

Study of fossil remains suggests that there is a link between Hyaenidae and Viverridae, but the origin of the latter group remains unresolved.

All species of the family Viverridae are small or medium-sized mammals which are chiefly found in the warmer countries of the Old World, although there are representatives in some temperate regions as well as introduced species in Hawaii, the Antilles and North America. In general all these animals are characterised by a streamlined body, elongated head, pointed muzzle, short legs and long, hairy tail.

The various members of this family prefer to feed on prey which they have themselves killed, but some of them also consume carrion. A few of them enjoy a mixed diet which comprises a good proportion of vegetable matter, especially fruit.

The typical dentition formula of the family is:

$$I: \frac{3}{3} \quad C: \frac{1}{1} \quad PM: \frac{4}{4} \quad M: \frac{2}{2}$$

the total number of teeth thus being 40. In some species, however, there may be a reduction in the number of premolars and sometimes molars too, as in the case of the fossa, an animal which looks something like a large cat and which has only 32 teeth.

The Viverridae display a wide range of coat colours and markings. Some are delicately spotted or exhibit a pattern of dark stripes, while others have large, irregular black patches against a light ground. Certain species are capable of aiming a jet of evil-smelling fluid from special glands in self-defence–as do the skunks of the family Mustelidae–and their livery is very spectacular and prominent, serving as it does as a warning signal of dissuasion.

Quite apart from this liquid which functions as a weapon, many Viverridae have a so-called 'scent sac' in the genito-anal region, with glands that secrete a characteristically strong-smelling substance.

Classification of this family is somewhat complicated by the fact that it comprises terrestrial, tree-dwelling and semi-aquatic species–the last represented by the water genets (*Osbornictis*), the marsh or water mongooses (*Atilax*) and the otter civets (*Cynogale*).

In 1945 George Gaylord Simpson divided the family into six subfamilies. The Viverrinae include the civets, at one time much sought after for their fluid secretions which were utilised in the perfume industry, as well as the genets and Asiatic linsangs. The latter belong to the genus *Prionodon* but there is also one rare African linsang, *Poiana richardsoni*.

Among the Paradoxurinae are the palm civets, while the Hemigalinae include the fanaloka (*Fossa fossa*) from Madagascar. On this island are also found members of the four genera forming the Galidictinae.

The largest subfamily, with the greatest number of genera, are the Herpestinae, including the mongooses of Africa and southern Asia and the ichneumon or Egyptian mongoose, found in southern Europe. The sixth subfamily, the Cryptoproctinae, comprise only a single species, the fossa (*Cryptoprocta ferox*), largest of Madagascar's carnivores. This animal, which is reputed though not proved to be a vicious killer, has a number of primitive characteristics suggesting an ancient line of descent.

CLASSIFICATION OF VIVERRIDAE

Subfamily	Genus
Viverrinae	*Poiana* *Genetta* *Viverricula* *Osbornictis* *Viverra* *Civettictis* *Prionodon*
Paradoxurinae	*Nandinia* *Arctogalidia* *Paradoxurus* *Paguma* *Macrogalidia* *Arctictis*
Hemigalinae	*Fossa* *Hemigalus* *Chrotogale* *Diplogale* *Cynogale* *Eupleres*
Galidictinae	*Galidia* *Galidictis* *Mungotictis* *Salanoia*
Herpestinae	*Suricata* *Herpestes* *Helogale* *Dologale* *Atilax* *Mungos* *Crossarchus* *Liberiictis* *Ichneumia* *Bdeogale* *Rhynchogale* *Cynictis* *Paracynictis*
Cryptoproctinae	*Cryptoprocta*

Facing page (above). The small-spotted genet is an unspecialised carnivore, adept at climbing and jumping, and thus very much at home in trees. The tail is normally slender but readily spreads in a 'bottle-brush', the genet's first indication of aggressive intent. (*Below*) The ichneumon or Egyptian mongoose does its hunting on the ground. Both these representatives of the Viverridae are found in the Mediterranean region and belong respectively to the subfamilies Viverrinae and Herpestinae.

CHAPTER 4

Game birds of the Mediterranean grasslands

Although many a mountainside in the Mediterranean region is densely covered with tall grass, shrubs and trees, this type of vegetation gives a false impression of the area as a whole, for these glimpses of the luxuriant forest which stretched across much of southern Europe at the dawn of history are nowadays few and far between. Even the most superficial examination will show that the countryside has been subjected to a gradual but implacable process of degeneration, four successive stages of which are visible in the maquis (areas covered by evergreen thickets but in parts given over to dry scrub), wasteland–the soil of which is thin and vegetation very sparse–steppe (partially grassland but mainly converted to pasture and cereal-growing), and desert. These varied types of vegetation bear witness to the changing face of the Mediterranean basin over the ages. The past and the future are there for all to see. If man continues to exploit the soil of the remaining areas of fertile land, the erosion process will be accelerated. If, by a miracle, he decides to reverse the trend, some parts can still be restored to their pristine state.

On bare ground the plants that grow first are those with a short life cycle, these gradually yielding place to others that take longer to develop and are more permanent. They all help to modify the composition of the soil and create the conditions necessary for a vegetation zone to acquire a greater degree of complexity and stability. The final phase in this ecological development is known scientifically as the climax, at which point the various plants, the subsoil in which they grow, the climate of the region and the animals that live there are all part of a stable, balanced environment which–provided there is no climatic change–will not in any way be modified.

The Mediterranean climax is generally described by botanists

Facing page : Vast stretches of former Mediterranean woodland have been transformed by centuries of human activity into farming country. Fields of golden corn conceal a wealth of animal life–including many birds perfectly adapted to such surroundings.

It is difficult to imagine that immense forests once covered areas which nowadays present a bleak spectacle of dry wasteland. Furthermore, scientists attribute this slow process of degeneration not to any drastic changes in climate but wholly to human interference.

as dry evergreen woodland, made up of tough, perennial foliage. Principal tree species are the small kermes oaks, evergreen holly oaks, cork oaks and conifers. These, together with other species such as wild olives, carobs and dwarf palms are variously distributed according to region, altitude and nature of soil. In many sandy coastal districts, for example, the most characteristic trees are Aleppo pines.

There are few places in the Mediterranean where this climactic stage still exists. Tree-felling, fires and overgrazing (particularly by goats) have shrunk the area of forest and transformed it into maquis. Although even in the latter zones there are still evergreen oaks, they are little larger than bushes (10–15 feet high at most). Interspersed with mastic, cistus and myrtle, this type of vegetational cover doubtless represents the climax in many Southern European regions.

If degeneration continues the maquis yields place–especially on chalky soil–to moor or wasteland, characterised by thyme and other aromatic plants; and should the destructive process not be halted there, the underlying rock may be exposed and herbaceous plants will predominate. After blooming briefly, these will eventually lose their aerial parts and the landscape will take on a desert-like appearance. The only surviving plant is the asphodel, the roots of which are capable of taking hold even on deeply eroded surfaces.

Once this stage has been reached it is next to impossible for attempts at reafforestation to succeed. The absence of tree cover simply allows heat, wind and drought to rule supreme. Each rain shower cuts more grooves into the hillsides, exposes another section of rock, loosens the soil and pushes it a little farther towards the sea. Soil erosion in many regions bordering the Mediterranean is so advanced that eminent ecologists say that it would take a thousand years for the terrain to be restored to its original state, and that such a transformation would be impossible in many areas even if the responsible authorities were to spend vast amounts of money and utilise the most sophisticated resources of modern technology.

Seas of corn

It has been said that in Roman times a squirrel would have been able to travel from Gibraltar to the Pyrenees without once coming down from the trees; but it is not strictly true to imply that the Mediterranean lands were completely covered by woods and forests. The discovery in Italy of fossil remains of bustards–typical birds of open plains–dating back beyond this period suggests that there must always have been areas devoid of tree cover. It was certainly in one such enormous stretch of grassland in the Middle East that man first learned to cultivate the soil and discovered the nutritive qualities of cereals, thereby transforming himself from shepherd to farmer. Yet this very discovery was to prove catastrophic because of the primitive, untutored agricultural methods employed. The soil, which was not properly turned over or enriched by fertilising agents, was rapidly exhausted. The celebrated 'fertile crescent' fringing the plains of Mesopotamia,

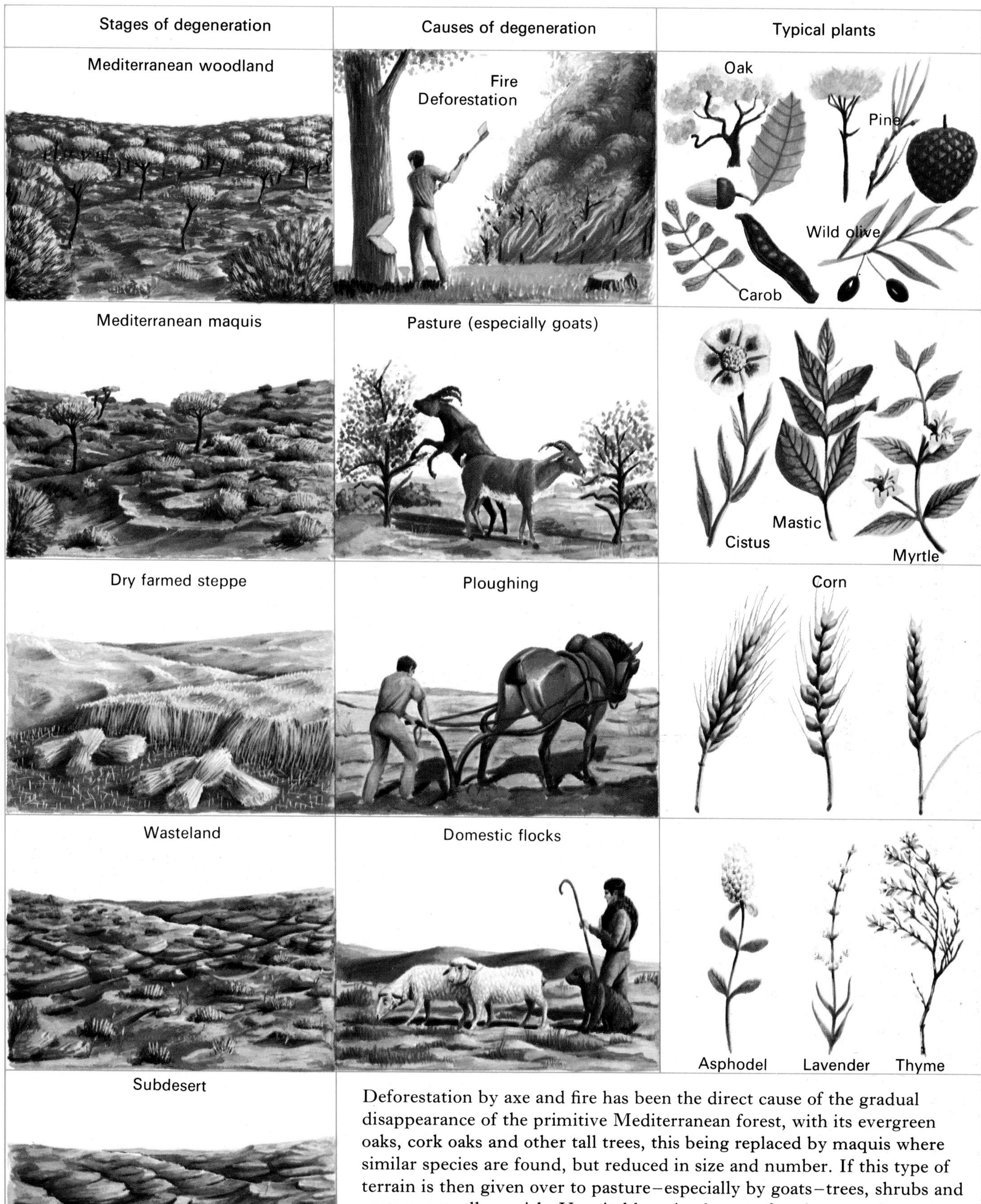

Deforestation by axe and fire has been the direct cause of the gradual disappearance of the primitive Mediterranean forest, with its evergreen oaks, cork oaks and other tall trees, this being replaced by maquis where similar species are found, but reduced in size and number. If this type of terrain is then given over to pasture–especially by goats–trees, shrubs and grass eventually vanish. Unsuitable or inadequate farming experiments subsequently lead to soil erosion, rendering it useless for cultivation. Practically devoid of plants, except for a handful of hardy species, the sterile wasteland is whipped by wind and rain until the underlying bare rock is exposed. The former forest has now been reduced almost to desert.

Geographical distribution of the great bustard.

GREAT BUSTARD
(Otis tarda)

Class: Aves
Order: Gruiformes
Family: Otididae
Length: male 40 inches (102 cm)
female 30 inches (76 cm)
Wing-length: male 24–27½ inches (61–70 cm)
female 19½–26 inches (49–66 cm)
Weight: male 15½–37½ lb (7–17 kg)
female 9–15½ lb (4–7 kg)
Diet: omnivorous
Number of eggs: 1–3
Incubation: 4 weeks

Adults
The male has a white collar and moustachial stripes. The lower parts are also pure white with rich chestnut breast. Head and neck are grey; back, subcaudals and small wing coverts are reddish with wavy black markings. The large wing feathers are brownish-black with white at bases of inner webs. The female is similar but much smaller and without moustachial stripes or dark brown markings on chest.

Young
Whitish when born, striped and spotted with ochre or brown. There are two black bands on the head and the throat is brown.

Facing page: Typical Mediterranean landscapes, representing two stages of the degeneration of an environment as a result of man's interference with the balance of nature. The original woodland has been replaced (*above*) by steppe partially given over to cereal growing. Incompetent cultivation has subsequently caused much of this land to become sterile (*below*) and the erosive effects of wind and rain are slowly turning it into desert.

where agriculture is assumed to have originated in the 5th century B.C., was speedily turned into semi-desert.

Most authors are inclined to attribute this process of deterioration to the primitive form of drainage practised by the earliest farmers. The numerous ditches they dug were much too shallow to permit thorough drainage so that the land was continually flooded. As the water table rose the saline content of the subsoil in these regions increased steadily so that in the end corn could no longer be grown. The sturdier barley crop provided an adequate substitute for a time, but eventually this too was superseded by handfuls of halophilous (salt-water) plants of no nutritional value. Thus within a few centuries the once-fertile land had gone utterly to waste. The same phenomenon, though not necessarily for the same reasons, occurred elsewhere, with the identical end result. This is what must have happened to Carthage, for example, once described as the granary of Rome.

Farmers in the Mediterranean countries now appreciate the importance of trying to check this slow deterioration of the soil. Although some soil is so poor that it will never yield more than one crop every six years, plants have been re-established with varying degrees of success in some areas and dry farming of cereals practised wherever soil has proved suitable. These cultivated zones have no natural equivalent for they are usually given over to one crop or at best a handful of species, in contrast to the enormous diversity found in the wild. There is some similarity, however, between these corn-growing districts and the open steppe with its grass cover, although in the latter there are of course no areas of fallow ground.

The conversion of large expanses of steppe to dry farming has affected local wildlife. Some species have disappeared, others have adapted to their new surroundings. In the Mediterranean basin, with its characteristic fields of hard wheat, the apparent monotony of the landscape may blind the casual observer to the abundant animal life of the region. The hare finds refuge among the ploughed furrows and the great bustard proceeds with majestic tread through the dusty fields when harvesting is done. Indeed the profusion of bird life is astonishing. The little bustard, partridge, quail, pin-tailed sand-grouse, black-bellied sand-grouse, peregrine falcon and kestrel are only a few of the species concealed in this sea of gold that stretches to the horizon.

The stately great bustard

The great bustard (*Otis tarda*) is the largest bird of the Mediterranean region, commonly found on open grassland and in cornfields. It is a shy bird, difficult to approach, and best observed either at dawn or at dusk when it ventures out for food. It may be seen alone or in small groups–depending on the season–but almost always in an exposed place with a free view of the surroundings, so that it can readily see potential enemies. Although quite capable of flying, the great bustard, like many other birds typical of this habitat, prefers to move about on the ground. It is never seen perching on a tree branch.

This impressive bird has a heavy body supported by powerful

The great bustard is ideally suited to life on the open plains and in cornfields. Very shy by nature, the bird possesses excellent vision and cannot easily be approached. The European population is sedentary, in contrast to the Asiatic, which migrates southward for the winter.

legs. The neck is thick and comparatively long, the head rounded, the bill short and broad. Considering the bird's size (the largest individuals may be just over 3 feet tall and weigh nearly 40 lb), it may seem surprising that it should habitually seek refuge and security on flat ground without obvious natural cover. But the astonishing thing about this species is its habit of hiding at the first sign of danger and the impression it gives, as many an ornithologist will confirm, of being virtually swallowed up by the ground.

Thanks to the cryptic coloration of its plumage, the great bustard is so well concealed among ripe stalks of corn or newly ploughed furrows that it is almost impossible to detect it, even when close at hand. In addition to blending with the surroundings, the bird, when alarmed, remains quite motionless.

Only in situations of extreme danger will the bustard attempt to escape by taking wing. It then straightens up, runs a few yards, generally in the direction the wind is blowing, and finally rises from the ground. It can work up a speed of about 30 miles per hour but normally flies at low altitude, skirting trees and hills. Although the species has no recognisable group flight pattern, the individual bird can easily be identified by its size, its slow, deliberate wing beats, the dazzling white on the underside of each wing and the enormous span–almost 6 feet–when both wings are fully extended.

Bustards on parade

Under a sky which is still dull with late winter greyness, as the dry, cold north wind sweeps across the plains where tender shoots of corn give promise of approaching spring, a flock of great bustards are assembled for their courtship ritual.

At break of day the elegant males are to be seen on a hilltop or in the middle of a field, decked out in all their nuptial finery. The wing feathers, extended like the fingers of a hand, drag along the ground, the tail is held stiffly open in the shape of a fan and the shoulder feathers are directed forwards. The head is tilted to the rear, almost flat against the back, the feathers of the moustachial stripe are erect and the sacs at the base of the neck are swollen to their fullest extent and slightly drooping.

Hardly recognisable in this seasonal array, the dominant males strut back and forth across their parade ground (sites are usually 5–8 miles distant from one another), awaiting the arrival of their prospective partners. The latter eventually make their appearance, advancing slowly in single file, causing a renewed flurry of activity among their suitors and even an occasional fight. This excitement generally takes place very early in the morning although nuptial displays are sometimes seen late in the evening and–more rarely–in the middle of the day.

After the pairs have coupled the females take themselves off to find conveniently sheltered spots within the large stretch of territory controlled by the dominant males. The older birds seem to return regularly to favourite sites–perhaps a vineyard or a quiet, isolated piece of grassland unfrequented either by man or his domestic animals.

Provided they are not disturbed, the birds come back every year to the same districts. Each female uses her feet to scoop a shallow hole in the ground, shaping it by lying in the hollow and moving her body to and fro. Every blade of grass is carefully removed from the vicinity with her bill. In this depression she lays from one to three eggs (generally two), the shells of which are brownish-green with irregular red or brown blotches.

After coupling the males form small groups and wander vaguely over the plain. These bands include birds that have not taken any part in the breeding activity and sometimes females which have, for one reason or another, lost their eggs. As a rule, however, the groups are exclusively male.

Meanwhile the majority of the females are installed in their nests and busy incubating. They are perpetually on the alert, ready to distract the attention of any intruder venturing too close. In the event of real danger threatening, they create a diversion by running off, tail stiff and wings drooping, deliberately drawing the predator away from the nest. This is the well known distraction display. Provided this ruse succeeds, they then resume their normal appearance and gait, walking with extreme caution or making a few short, exploratory flights, finally going back to the nest to resume incubation.

The eggs hatch after four weeks and the fledglings, which are not very mobile, remain near the nest for some days. At this stage complete immobility is their only means of defence. They flatten themselves against the ground, not moving a muscle, until such time as the cause of their alarm has been removed and they recover their sense of security.

Young bustards are very partial to ants and in certain regions the birds' nesting sites are always found in places where there are plenty of anthills. Shortly after birth, the mother instructs her fledglings in the art of catching the insects by scratching away with their claws at the mounds to dislodge the occupants.

Initially each family of bustards lives apart from the nest, but as the youngsters grow and their instincts for self preservation increase, there is a tendency for the various groups to edge closer together so that fairly large colonies form. These do not disperse until the following spring.

The great bustards inhabiting the Mediterranean region are sedentary by habit and only embark on short journeys in search of food. Those living on the southern steppes of the Soviet Union have the same characteristics but those found in northern countries are for the most part migratory. As early as August a few individuals, usually the younger birds, have left for the breeding grounds and are heading south. Flocks may comprise as few as thirty or forty or as many as two hundred birds. The older birds seem to be rather more reluctant to move and some of them leave it to the very last moment, when the first snows appear. The migrating bustards fly at low altitudes, choosing the wildest type of open terrain for their temporary halts. The return journeys in spring are more predictable in their timing, most of the birds appearing, alone or in pairs, between mid-March and mid-April.

Great bustards are noted for their extraordinarily wide range of food. Birds, mice, lizards and frogs are the more common forms

Male great bustard in breeding plumage

Female great bustard

of prey, yet examination of their stomach contents has even revealed stones. The diet, however, will vary to some extent with the season. In spring and autumn the birds are content to live mainly on grass and seeds whereas in summer they also consume large numbers of beetles, grasshoppers and small vertebrates.

Unable to survive without drinking regularly, the bustards normally roost as close as possible to water. Many districts which would appear to be ideally suitable as far as vegetation goes are avoided if there are no ponds or streams in the vicinity. This explains why the species is never found anywhere near a desert and the range of distribution is further restricted to open ground with good grass or shrub cover, befitting their retiring nature and comparative lack of agility in the air.

The little bustard

The great bustard, whether found in Eurasia or Africa, is immediately identifiable by its size, weight and low, somewhat lumbering flight. In some Mediterranean countries it is simply nicknamed the 'slow bird'. How disconcerting, therefore, to see a group of these birds suddenly rising with a great flutter of wings and a strange whistling noise from the middle of a cornfield, soaring to a considerable height in a matter of only a few seconds. This graceful performance, so uncharacteristic of a heavy ground bird, has a simple explanation, for in fact we have been watching a different species–the little bustard (*Otis tetrax*). Our momentary error is understandable since we have probably received only a fleeting impression and at a superficial glance this bird appears to be a smaller version of the great bustard.

Granted that the little bustard is less than half the size of its more imposing relative, it is nevertheless astonishing that it should be so powerful, fast and manoeuvrable in the air, considering that the great bustard and other members of the family are comparatively slow and inept. Yet the explanation is not all that mysterious. The flying prowess of the little bustard has to a large extent been dictated by its enemies. The great bustard, weighing 15 lb or more, is normally too large and strong for a falcon; but a little bustard that weighs only 2 lb is ideal prey for those typical raptors of the steppes–peregrine, saker and lanner falcons. If it ventures too close to woods it may be picked off by a goshawk; in hilly terrain it risks being attacked by a Bonelli's eagle, and on the shores of lakes or rivers it is quite likely to fall victim to a marsh harrier.

In view of these diverse predatory hazards it may seem strange that the little bustard should deliberately choose to live, like its less vulnerable relative, on open plains lacking opportunities of concealment. But if one carefully studies a group of these birds searching for food among the stubble, pecking at the ground for seeds or suddenly striking out at a grasshopper, it is evident that their confident manner stems from the functioning of an efficient security system. Sentinels posted here and there keep an unwavering watch on the distant horizon. Standing a foot and a half tall, long necks stretched out to give added height, these birds on guard duty are able to spot a predator when it is still some miles

Facing page : The great bustard has adapted perfectly to its characteristic habitat of grass-covered and cultivated steppe. Its wide range of food includes seeds, rodents, birds, reptiles and amphibians–consistent with its ground-dwelling activities. The cryptic coloration of its plumage–whether against a background of recently ploughed land or fields of ripe corn–conceals its presence from predators.

The characteristic whistling sound made by a flock of little bustards is associated with the wing structure of the males. The fourth wing feather is shorter than the others so that the air 'whistles' through the adjacent gaps. The noise cannot be heard after the feathers moult. In the female the remiges are all of the same length.

The little bustard is almost a small replica of the great bustard, distinguished mainly by its flapping flight, contrasting with the slow beating of the great bustard's wings.

away, affording more than enough time to warn their feeding companions and allow them to take evasive action. If the birds elect to fly away, they all take off vertically with powerful wing beats, wheeling up and round in a wide semi-circle. The characteristic high-pitched sound of the flock in flight is produced by the wing feathers of the males. The fourth feather is shorter and narrower than the others and the air literally 'whistles' through the gap formed between this and the adjacent remiges. Thus the action of flying is in itself a sufficient alarm signal. One aroused male can in this way alert all those birds which are still unconcernedly pecking away at the ground and provoke a general exodus. Furthermore, when the birds migrate, the noise made by this concerted wing beating serves as a rallying call to those lagging behind, urging all the family groups to band together into one enormous flock.

Life-saving stratagems

Taking to the air when threatened by an enemy may, however, be only a final resort. The sight of a peregrine falcon hovering overhead is more likely to produce a very different reaction whereby each bird simply 'freezes' and does its best to disguise its natural outlines. The cryptic colours of the plumage, similar to those of the great bustard, render the small bird almost invisible from above. But even if the falcon succeeds in flushing a group of little bustards from cover, the latter have a number of subtle ruses which may yet save them from the terrible claws of the foe. One individual, often the oldest and most experienced of the group, will take wing; its objective, however, is not to escape but to attract the predator's attention. The normal reaction of the falcon is to dive directly at the enticing prey below, but when its talons are within yards of the target the latter insolently drops down into the grass and vanishes. This is the signal for the rest of the flock to rise from the ground and fly off in the opposite direction, generally into the wind. The falcon is handicapped by its inability to manoeuvre rapidly at low level and has lost its advantages of surprise and speed. Predators often have to work hard for a living!

Even if a flock of feeding or migrating little bustards should be taken unawares by a falcon swooping down from a height of several thousand feet, they still have a good chance of using similar tactics to save themselves. Felix Rodriguez de la Fuente has for many years hunted little bustards with trained falcons, especially in the vicinity of airfields where the birds pose some danger to aircraft as they are sucked up by the jet engines. He has testified to the extraordinary agility of the bird and the way in which it will escape from its pursuer's clutches at the last possible moment.

The raptor's best chance of success is when it is gliding at a height of about 1,000 feet directly above its prey. Once the victim is clearly sighted the falcon folds its wings and launches an attack, diving perpendicularly or at an oblique angle at a speed of more than 125 miles per hour. The bustard will probably have had enough time to get off the ground and gain some altitude, in which case the predator checks its fall at the point where it is

The little bustard is hunted by a large number of predators, including falcons. A peregrine falcon will swoop down from a height and launch an attack from below and behind (1). Although its chances would appear to be poor, the little bustard may in this situation have recourse to a stratagem, which is basically instinctive. Just as the predator prepares to seize its victim, the latter ejects a sticky stream of excrement, temporarily blinding the falcon and soiling its plumage (2).

LITTLE BUSTARD
(Otis tetrax)

Class: Aves
Order: Gruiformes
Family: Otididae
Length: 18¼ inches (46 cm)
Wing-length: male 10–11 inches (25–27·5 cm)
female 9¾–10¾ inches (24·7–27 cm)
Diet: omnivorous
Number of eggs: 3–4
Incubation: 20–21 days

Adults
Male's face is blue-grey, upper part of head and back sandy buff with wavy black markings; lower parts white. In the breeding season the neck is black with two white bands. The female's back is paler, with less prominent black markings; her lower parts are not as gleaming white and there are no bluish tints on the face.

Young
Covered with white or pale yellow down at birth, with brown stripes and spots on the back.

slightly below and to the rear of its intended victim. It then swoops upwards once more so that it can seize its prey from behind. But just as the curved talons stretch out, so near that they are almost touching the smaller bird's body, the latter unexpectedly ejects a stream of sticky green excrement, temporarily blinding the assailant – a last-ditch defence mechanism which frequently proves to be a life-saver. The accuracy with which the bird directs its droppings at the luckless falcon is quite remarkable, the plumage of the raptor usually being completely fouled, thus preventing further offensive action.

Defecation and urination caused by the relaxation of the sphincter muscles under the stress of acute fear is a common phenomenon in many animals. The transformation of an instinctive action into a defensive weapon, as in this case, is an adaptive process brought about by the pressure of enemies, for there is no other habitat where such a vast concentration of falcons of different species is to be found.

Thanks to natural camouflage and such ingenious and unexpected stratagems of self-protection, the little bustard holds its own against enemies. But if need be the bird can make itself easily recognisable by means of the conspicuous pattern of the wings. The wing feathers are pure white and the tips bordered with black. This vivid colour contrast, as well as the curvature of the wings, enables the birds to attract one another's attention at a distance of several miles, especially in the course of nuptial flights in dazzling sunlight.

As with many other birds, the colour of the plumage varies from season to season. Even the characteristic whistling of the reproductive males is lost when the fourth primary remex moults. During the long autumn and winter months it is difficult to distinguish the two sexes from a distance for their silhouettes are similar and they are garbed in matching colours – white below, brown with narrow stripes above. Only very close examination shows that the wavy lines on the back are much fainter in the male than in the female. In summer, on the other hand, the male's plumage is extremely spectacular, his neck adorned with a pure black band, the back and the head taking on a sandy hue and the face turning a magnificent blue-grey.

This is the season when the males come together, like the great bustards, for their nuptial performance. In regions where little bustards are especially numerous the arenas chosen for the purpose are heavily trampled and identifiable at all seasons because of their lack of grass.

The females excavate a hole which is a couple of inches deep and 6–7 inches in diameter, laying 3–4 eggs. When incubating, the white belly is completely covered by the depression so that only the upper part of the body is visible, effectively concealing the bird from predators. But if an enemy should venture uncomfortably close, she abandons the nest and makes off as fast as she can through the grass. When sufficiently far away she flies up a few feet in an attempt to draw the predator off in the wrong direction. In a real emergency, when the incubating female is caught by surprise and has no time to employ such diversionary tactics, she spills her excrement over the eggs and abandons them,

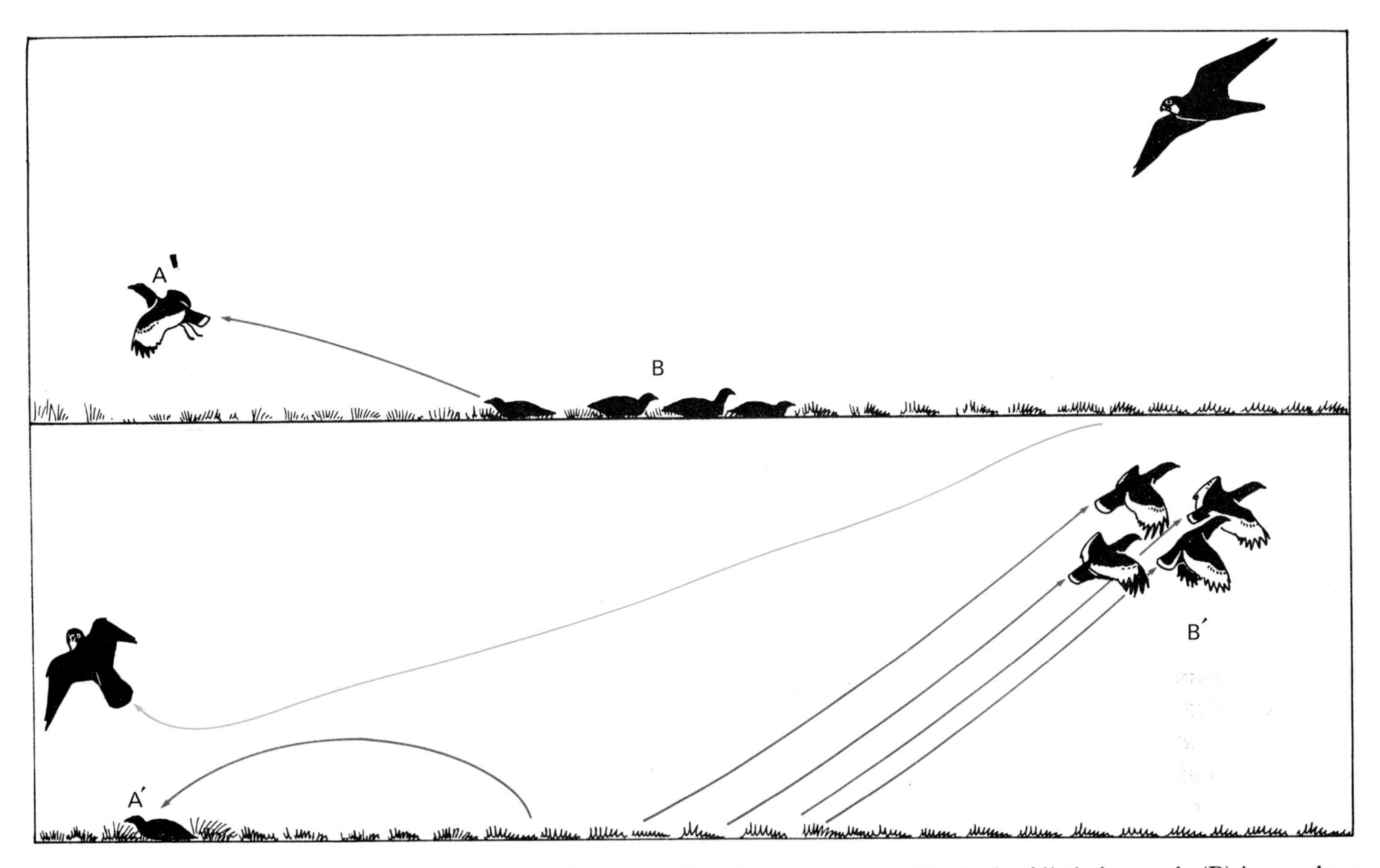

If a flock of little bustards (B) is caught unawares on the ground by a falcon the plan of action is for one bird, generally the oldest, to take wing (A) and divert the raptor. As the falcon dives the bustard drops down into the grass and disappears from view (A′). Meanwhile its companions soar off to safety in the opposite direction (B′).

never to return. There is a perfectly good reason for this apparently pointless exercise, for the predator, remembering the nesting site, will perhaps venture back in due course, only to find the prey has seemingly vanished.

Incubation lasts 20–21 days and is undertaken by the female alone. Her partner meanwhile remains nearby, mounting guard on a hillock, mound or ridge.

Soon after they hatch, the fledglings accompany their mother in search of beetles and grasshoppers. At the first sign of danger they hide in the grass, flattening their bodies against the ground, necks tautly stretched.

By the end of August the fledglings are already capable of flying and join together in bands, crossing and recrossing the plains in short test flights, preparatory to facing the great challenge of the autumn migration.

About a month later–the interval varies from one region to another–the birds become increasingly gregarious and groups may consist of up to one hundred individuals. Smaller groups comprising only two or three little bustards are found in areas where the species is rare. Winter quarters are in southern Europe, Africa, Asia Minor, southern Asia and north-eastern India. When large flocks are involved brief halts are made at intervals, provided the surroundings are suitable.

Flocks of little bustards arrive in northern Spain from the end of September or the beginning of October, where they mingle with sedentary colonies whilst awaiting the first seasonal frosts. Then they move on to the south, to Estremadura, Andalusia and the African continent–where the climate is warmer–making their way back to Europe in the spring.

Geographical distribution of the little bustard.

The little bustard lays her eggs in a shallow depression in the ground which she has scooped out and shaped by randomly shifting her body to and fro. When she is incubating only her speckled back projects above ground level and she is well hidden from hovering predators.

The red-legged partridge is a popular game bird, abundant in the south-west Mediterranean region where attempts have been made to protect it by destroying many of its natural predators.

The red-legged partridge

Much more abundant in the Mediterranean region than either the great or little bustard, and considerably less retiring, is the red-legged partridge (*Alectoris rufa*).

This plump bird has a restricted range in comparison with that of the better known common partridge (*Perdix perdix*), breeding in Spain, Portugal, the south of France and in southern and eastern counties of England. It is hunted by many carnivores and birds of prey, and because of its remarkably sustained and rapid flight is a highly popular game bird. For this reason man has done his best to ensure that the species should continue to multiply. One obvious way to achieve this has been to kill off many of the birds' natural predators; but, as has proved to be the case in similar situations elsewhere, this 'solution' could hardly have been more short-sighted. It completely overlooks the elementary laws of ecology which demand, in order to maintain natural equilibrium, that there should be a continuous interaction of hunters and prey. Waging war on known predators such as vultures, kites, imperial eagles, Bonelli's eagles, buzzards, lynxes and foxes, failed to have the desired effect, partly because it was not realised that these carnivores were not by any means exclusively dependent on the red-legged partridges which were to be safeguarded. Careful studies of their feeding habits would have shown – as became evident at a later stage – that the hunters in question actually killed a relatively small number of partridges and that they preyed just as frequently on members of the crow family – birds that are potentially harmful in the sense that they destroy the nests of other species. Although the Corvidae do not cause too much damage as long as their population is kept under control by predators, the sudden removal of the latter is bound to cause serious local problems. Thus coincidentally with campaigns against the carnivorous mammals and birds of prey of the Mediterranean region there was a spectacular increase in the numbers of crows, magpies and jays.

In south-western Europe, particularly in Spain, the loud, harsh cry of the red-legged partridge can be heard as early as February when the days are already warm and cloudless although, in terms of the calendar, spring has not yet officially arrived. This characteristic call of the male partridge, announcing his territorial claims and denying his rivals the right to intrude, is difficult to describe in words but has been likened to the wheezing of a labouring steam engine. If another male of the species should choose to ignore the warning and try to engage in a poaching exercise to enlarge his own territory, a noisy encounter will follow in which each combatant shows off his vocal powers as evidence of his potential fighting prowess. As often happens in such situations, no matter what kind of animal is involved, the original owner of territory usually emerges the winner. Once freed of any menace from his rival the victor sings even more loudly as if indulging in a hymn of triumph, to which the females respond by drawing near.

Only if the intruder is particularly stubborn and powerful will there be a fight proper. The two birds then face each other, heads

erect and slightly tilted to show off the brilliant colours of the crop, necks inflated and wings half open and drooping. Now and then they peck at the ground as if in search of imaginary food. If neither is prepared to give way the rivals now join battle, lashing out with bills and claws but seldom inflicting any serious injuries. Eventually the bird which is clearly losing withdraws from the field, head hanging low. The victor immediately strikes up his territorial challenge with renewed vigour, this time with the object of attracting a female. The ensuing courtship display begins with the same rather aggressive movements that previously characterised the combat ritual. Finally the male approaches his partner from behind, stretching his neck and swinging his head up and down, grasps her firmly and mounts her.

After the birds have mated the male selects several sites, strategically situated at different points in his territory, only one of which will eventually be used as a nest. He burrows a shallow hole in the ground, stooping slightly and turning around in circles, giving out soft cries to encourage his mate to join him. He then settles down in the cavity, shifting his body about to give the nest its final shape. Finally he lines the hole with leaves, twigs, grass and any other material that happens to be lying nearby. Having completed his work, he proceeds to the next site and starts again, employing identical methods, the female being content

In the breeding season, red-legged partridges live in pairs, the male first staking out the imaginary frontiers of his territory and defending it noisily against rivals.

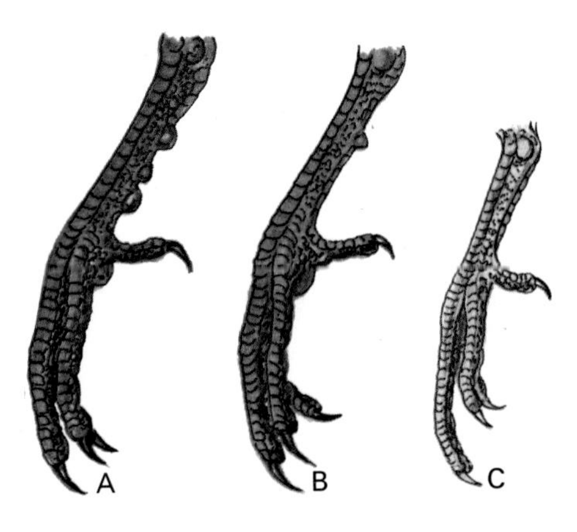

The age and sex of a red-legged partridge can be determined by the number of spurs. The adult male (A) has two or three, the adult female (B) only one. There is no sign of a spur in the young (C).

Geographical distribution of the red-legged partridge.

to be a spectator or perhaps to try the nest for size and comfort when it is finished.

Eventually the female lays 10–15 eggs in one of the nests, after which her partner pays no more attention to the other sites and allows them to deteriorate. Only when the last of the eggs is laid does incubation commence. Incidentally, the English ornithologist Derek Goodwin, who has devoted much time to the study of the behaviour of this species, points out that red-legged partridges in captivity have been observed to deposit eggs in two different nests, one clutch being incubated by the female, the other by the male. This does not appear to occur in the wild, although it occasionally happens that two females may lay their eggs in the same nest. In any event, the male never assists in incubating them. For a while he remains close by to guard the nest but later wanders away to join other males, leaving the female to rear the fledglings alone.

Incubation lasts about 25 days, during which period the female seldom abandons the nest, except for brief intervals in order to feed, drink or take a dust or sand bath. The newly hatched fledglings also indulge freely in this last activity, choosing an empty patch of ground, ruffling their feathers and then using bill, feet and wings to smother the entire body with dust. They extend the neck and rub it vigorously against the chest. When this pastime is over they get to their feet and shake themselves clean.

As soon as her fledglings hatch the mother emits gentle cries of recognition, using the same signals whenever she subsequently wishes to gather them to her side. As with all nidifugous species, the baby partridges are able to stand upright and, although initially somewhat unsteady, are equally capable of walking. Within two days of birth they are running about after their mother. When one of the fledglings discovers food it imitates her soft call to summon its siblings. At this stage they eat various insects–particularly ants–whereas the adults are basically vegetarian. Only in the summer will the latter supplement their grass and seed diet with insects, spiders and other invertebrates.

When they have eaten their fill the fledglings hurry back to seek maternal warmth, slipping underneath the mother's body, and even getting up on their tiny feet to make closer contact. This is all very well when they are small but having grown to a moderate size this is no longer tolerable, for the simultaneous action of rising simply lifts the mother right off the ground and they have to make alternative arrangements.

While they are comparatively small the mother settles her offspring for the night in a sheltered spot and protects them with her body. Later, when they are capable of flying, the youngsters sleep in the open but continue to huddle against one another as best they can to keep warm.

Although the family may initially consist of a mother and her 10–12 young, the mortality rate is so high in this species that numbers rapidly decline during the first summer. By September there may be no more than 2–3 survivors. These family groups then come together for mutual protection and stay united until spring, when the birds disperse for breeding activities.

RED-LEGGED PARTRIDGE
(Alectoris rufa)

Class: Aves
Order: Galliformes
Family: Phasianidae
Length: 13½ inches (34 cm)
Wing-length: male 6¼–6½ inches (15·6–16·5 cm)
female 5¾–6¼ inches (14·7–16 cm)
Weight: up to 1¼ lb (550 g)
Diet: grass, also insects in summer
Number of eggs: 10–15
Incubation: 25 days

Adults
Bill and feet red, black-bordered white gorget, white brows, greyish flanks stippled with black, white and chestnut markings, blue-grey chest. Female's colour less prominent.

Young
Greyish or chestnut marks on back when born, dark stripes on wings and flanks. Lower parts reddish-grey, feet orange.

A confrontation between two rival male red-legged partridges begins with both birds emitting their strange wheezing cries, designed to have a deterrent effect (1). If the aggressor fails to heed this warning the birds circle each other, neck inflated and one wing drooping (2). This mutual show of strength is interrupted now and then as the birds peck at the ground (3), until one of the contestants launches an attack (4) by grasping the other's back and trying–usually without success–to damage its skull with powerful blows of the beak. The second bird is generally able to disengage and make its escape. Numerous studies of this species (*lower*) have revealed other interesting features of behaviour and typical attitudes in different situations. If the bird sees an object which excites fear or curiosity it approaches with great caution (A); and when it runs away it darts a glance out of the corner of the eye (B) to see what is happening behind. The bird often sleeps on its feet (C). The female gives out soft cries to assemble her fledglings (D) and protects them with the warmth of her body at night (E).

The migratory quail

Many authors describe the quail (*Coturnix coturnix*) as a partridge in miniature but in fact there are several distinguishing features, both in appearance and behaviour. The quail is no larger than a blackbird and its plumage contains a broader range of colours–brown, yellow, white and grey–than that of the partridge. The bill of the quail is shorter, there are no spurs on the legs and there is a greater difference in appearance between the sexes (the male and female partridges are virtually alike).

Apart from these outward distinctions there are even more important variations in habit. Although the quail is also a popular game bird it is the only one in Europe to migrate, travelling thousands of miles from breeding ground to winter quarters. This may not seem especially remarkable, for there are plenty of birds that undertake the same complex, arduous seasonal journeys. The quail, however, is a member of the Galliformes–birds that are not customarily renowned for long migrations. This is why ornithologists are so keenly interested in the seasonal movements of flocks of quails, ringing the birds as a means of tracing their flight patterns and migration routes.

There are two principal quail populations, one in southern Africa, the other in the Palearctic region. The former is sedentary, the latter only partially migratory. Birds living in southern Europe and Asia (and these include the inhabitants of the Iberian peninsula, notably on the west coast in Estremadura and in Andalusia) remain there all year round. It is the quail colonies living farther north which are the true migrants. These fly south over Spain and Italy between the months of August and October and return by the same routes between March and May of the following year. The birds show remarkable powers of endurance, for many of them make long uninterrupted journeys across seas or fly high over mountain ranges. It is a strange fact that in the spring the last to return are the females and young. At this season many quails that have wintered in Africa head for the Mediterranean lands, alighting in vast numbers on the open plains where they mingle with the bustards and partridges. Others continue into central Europe and breed there.

The quails' breeding season is characterised by a great deal of noise and activity. As in the case of the red-legged partridge, the male defends his nuptial territory against all comers but does not necessarily confine himself to a single partner. He may be monogamous or polygamous, depending on how many females are available in his particular area. Unlike the red-legged partridge, he plays no part whatsoever in finding a site for the nest or building it, and he emulates the latter in assuming no responsibility for incubating the eggs or, needless to add, for rearing and protecting the fledglings.

The mother-to-be therefore has to assume all the duties that are in many species shared by both parents. She lays 7–12 eggs–yellowish, speckled with brown–and subsequently incubates them for approximately three weeks. The fledglings are nidifugous and show the characteristic rapid growth rate and precocity. At six weeks of age they are able to fly.

Geographical distribution of the quail.

QUAIL
(Coturnix coturnix)

Class: Aves
Order: Galliformes
Family: Phasianidae
Length: 7 inches (18 cm)
Diet: herbivorous
Number of eggs: 7–12
Incubation: 21 days

Similar to partridge but much smaller. The sandy-coloured plumage has brown and black patches and longitudinal white stripes. Brows are creamy and there is a creamy band on the head. The male has black marks on the neck; in the female these are yellow. The chest of the female is striped. The young look similar to the females.

Facing page: From March onwards the migratory quails of the central parts of Eurasia return from their winter quarters to breed. The female lays 7–12 speckled eggs in a rudimentary nest consisting of a hole in the ground, lined with scraps of vegetation.

ORDER: Galliformes

The order Galliformes comprises fowl-like birds of medium or large size. They have a comparatively heavy body, short, broad wings and a downward-curving bill. The sturdy feet are furnished with four toes, three of which point forwards. The fourth, which is fairly long and raised above the other three, is directed to the rear.

Galliformes are as a rule ground birds, although some species live in trees. Many of them are omnivores.

The majority of these birds are sedentary, occasionally making short journeys in quest of food. Only the quails – and not all of these – are genuinely migratory.

According to the most commonly accepted classification there are two sub-orders – Galli and Opisthocomi. The former comprises six families and some 240 species; the latter one family with a single extant species.

Of the first suborder the Megapodiidae are essentially ground birds, characterised by their rather heavy, lumbering flight. They are found in Australia, New Guinea and a number of small Pacific islands. One highly distinctive feature of such birds is the unusual manner in which they incubate their eggs – not, as in the case of most other species, by warming them under the body, but by utilising the natural heat of the sun, of volcanic terrain or decaying vegetation. They are often known as mound builders because of their habit of burying their large eggs in a mound constructed of heaped earth and leaves. As this decomposes, the accumulated material releases sufficient internal heat for the eggs to develop.

The seven genera of Megapodiidae are *Macrocephalon, Eulipoa, Megapodius* (including three species of scrub fowl), *Alectura* (brush turkey), *Aepypodius, Talegalla* and *Leipoa.*

The family Cracidae comprises 38 species of medium-sized birds, represented in the New World from the southern parts of the U.S.A. down to Argentina. In all species the tail is comparatively long (consisting of twelve rectrices) and the head is sometimes adorned with a crest. Divided into eleven genera they are arboreal by habit, the best known being the slender, easily domesticated curassows and the rather smaller guans, all species being forest dwellers and markedly gregarious.

The Tetraonidae are widely distributed throughout the cold and temperate lands of the Holarctic region, most of them being ground birds. The largest members of the family are the six species of grouse, the smallest the four species of ptarmigans. The latter usually display near-white plumage in the winter, which turns grey or brown in the warmer seasons.

The Phasianidae include birds from both the Old and New Worlds – a large and varied group made up of 178 species. These include pheasants, peafowls, partridges, quails and francolins. Many of the pheasants are gloriously coloured, in contrast to the somewhat drab hues of many partridges and quails.

There are four genera and seven species of Numididae or guinea-fowls – *Numida, Guttera, Agelastes* and *Acryllium.* Ground birds of gregarious habit, their distribution is restricted to Africa south of the Sahara, Madagascar, the Comoro Islands and parts of Arabia.

The Meleagrididae are the turkeys – ground birds from the American continent. There are only two species, the wild turkey (*Meleagris gallopavo*) and the ocellated turkey (*Agriocharis ocellata*).

The only surviving representative of the Opisthocomidae (the single family belonging to the suborder Opisthocomi) is the curious hoatzin (*Opisthocomus hoazin*), found only in the hottest parts of South America. It is not an accomplished flier and lives in trees. The fledglings are born with claws at the tip of either wing, which disappear in a couple of weeks.

CLASSIFICATION OF GALLIFORMES

Suborder	Family
Galli	Megapodiidae Cracidae Tetraonidae Phasianidae Numididae Meleagrididae
Opisthocomi	Opisthocomidae

Facing page : The pheasant (*above*) and the turkey (*below*) are both representatives of the order Galliformes, the former belonging to the family Phasianidae, the latter to the Meleagrididae.

ORDER: Gruiformes

The order Gruiformes groups together birds that are essentially ground-dwellers. Some of them embark on an occasional flight, others never take to the air. The majority of them are seen on the banks of rivers and streams or in marshes and swamps. Certain species, notably those belonging to the family Heliornithidae and some representatives of the Rallidae, are completely adapted to an aquatic form of life.

The order is subdivided into eight suborders and twelve families.

The suborder Mesoenatides comprises only one family–Mesitornithidae, found in Madagascar. These birds construct a nest fairly close to the ground so that they can reach it without having to take wing.

The Turnicidae and Pedionomidae both belong to the suborder Turnices. The Turnicidae include fifteen species, one of which, the Andalusian hemipode (*Turnix sylvatica*) is found in Europe. The only member of the Pedionomidae, found on the Australian savannahs, is the collared hemipode (*Pedionomus torquatus*).

Four families make up the suborder Grues. The Gruidae, which include the cranes, are large birds which sometimes stand over 5 feet tall and have a wingspan exceeding 6 feet. The five genera–*Grus, Balearica, Anthropoides, Tetrapteryx* and *Bugeranus*–are distributed all over the world, apart from South America. The limpkin (*Aramus quarauna*), ranging from southern Georgia and Florida down to Argentina, is the only species of the Aramidae. The Psophiidae (trumpeters) comprise the single genus *Psophia*, with three species found exclusively in the Neotropical region (*Psophia crepitans, P. viridus* and *P. leucoptera*). Most of the members of the family Rallidae are aquatic birds, living all over the world with the exception of the Arctic and Antarctic regions; some of them, however, are found on arid plains. The most typical European representatives of the family are the rails, moorhens and coots.

The only family belonging to the suborder Heliornithes are the Heliornithidae–the finfoots or sun grebes. The three species are inhabitants of tropical and subtropical countries.

Sole representative of the Rhynochetidae–suborder Rhynocheti–is the kagu (*Rhynochaetus jubatus*), from the island of New Caledonia. This bird has a high erectile crest on its head and exceptionally keen vision.

The suborder Eurypygae, with its one family Eurypygidae, is represented by the sun bittern (*Eurypyga helias*), a pigeon-sized bird which owes its name to the splendid pattern and colour of its wings. It is found in woodland from Mexico southwards to Bolivia and Brazil.

The seriemas of the family Cariamidae belong to the suborder Cariamae. There are two species, the crested seriema (*Cariama cristata*) and Burmeister's seriema (*Chunga burmeisteri*), both of them living in the driest regions of South America. These birds feed on insects and vertebrates, dismembering them in the manner of birds of prey.

The suborder Otides consists of the single family Otididae. These are the bustards, essentially inhabitants of the plains and steppes of the Old World. The family is made up of eight genera and twenty-two species. Among the best known are the great bustard (*Otis tarda*), the little bustard (*Otis tetrax*) and the houbara bustard (*Chlamydotis undulata*), notable for their crests and feathered neck tufts.

The Gruiformes are of very primitive stock and many fossils have been found. The Rallidae, Gruidae, Cariamidae, Aramidae and Otididae certainly date back to the Tertiary. Some giant forms–Gastornithidae–lived in France at the beginning of the Tertiary, while the Phororhacidae were found in South America during the Pliocene. These were wingless carnivores, fleet of foot, and about the size of ostriches.

CLASSIFICATION OF GRUIFORMES

Suborder	Family
Mesoenatides	Mesitornithidae
Turnices	Turnicidae Pedionomidae
Grues	Gruidae Aramidae Psophiidae Rallidae
Heliornithes	Heliornithidae
Rhynocheti	Rhynochetidae
Eurypygae	Eurypygidae
Cariamae	Cariamidae
Otides	Otididae

Facing page: The order Gruiformes comprises twelve families which include birds of varying size and appearance. The family Gruidae are the cranes, among the most beautiful of which are the crowned Cranes (*above*). One of the representatives of the Rallidae is the purple gallinule (*below*).

CHAPTER 5

Winged invaders of field and orchard

Man's interference with nature has invariably had a profound impact on long-established animal communities. The hacking down of woods and forests, gradually destroying vegetation to the point that only a few sturdy plant species remained, had a double effect. Some animals disappeared from their traditional habitats, others multiplied alarmingly. The former, since they were no longer able to find their customary food, were prevented from breeding; the latter, freed from competition for living space, took advantage of the expansion of agriculture, having always fed on the types of seed and grain that man was now providing so generously in the form of cultivated crops. These species, with little to fear from predators, now reproduced at such a dizzy rate that they presented a threat to farming and the economy.

Thus the little Java sparrow (*Munia oryzivora*), feeding principally on rice, was subjected to a merciless campaign of extermination after experts produced figures showing that the flocks were consuming quantities of rice sufficient to feed a human population of two or three millions. Peasants armed with sticks and scarecrows kept watch on roofs of houses and climbed trees to prevent the birds from roosting; unable to gain footholds, the sparrows fell exhausted to the ground after several hours and were quickly dispatched. In the short term this operation had the expected result for most of the winged pillagers were destroyed. Some years later, however, the side-effects began to appear. In their enthusiasm the peasants had impartially slaughtered harmful and beneficial species alike and now the rice crop was at the mercy of hordes of insects. The situation was more critical than before.

It is admittedly difficult to place any species in a watertight category. A bird may be perfectly harmless at one period

Facing page : The carrion crow is one of the few species to have derived positive benefits from the changes which man has made to the traditional habitats of the Mediterranean region. The disappearance of woodland, together with many typical predators, followed by the cultivation of cereal crops, are the principal factors which have permitted these birds to multiply and thus pose a threat to farmers because they tend to take grain.

The starlings, which have a boldly speckled plumage in winter (*above*) and are more or less uniformly coloured in summer (*below*), are noted for their gregarious habits (*facing page*). This social organisation has enabled the species to spread far beyond its original European habitats to all five continents.

of the year and a veritable scourge at another, simply because of its seasonal food preferences. Thus the starling, when it winters in the Mediterranean, can bring ruin to olive growers by wiping out the entire crop in certain localities, whereas in central Europe it provides a service to farmers by feeding, during the breeding season, on vast numbers of destructive insects. Consequently peasants in the latter areas have traditionally protected starlings' nests, although this is less frequent nowadays in view of the damage the birds cause elsewhere.

The ubiquitous starling

The starling (*Sturnus vulgaris*) has a wider world distribution than any other bird species, this being partly due to human assistance but principally to its own remarkable adaptability. Originally confined to the Palearctic region, it subsequently invaded Iceland, the Azores and other islands in the North Atlantic, North and South America, New Zealand, Australia and South Africa. The bird had always been popular in Europe and those responsible for introducing the species abroad–with the best of intentions–never imagined that it would be capable of multiplying to plague-like proportions. Thus in the 19th century someone had the pleasant idea of introducing into the United States all the birds mentioned in Shakespeare's plays, which involved releasing a small group of starlings in New York's Central Park. It did not take the newcomers long to spread across the country, up into Canada and down into Mexico. There are now some five hundred million starlings in the New World.

In central and northern Europe starlings have traditionally lived in dense forests or on grassland with scattered trees and shrubs. Towards the end of March each male installs himself in a natural cavity or a hole drilled in a tree by a woodpecker, a cleft in a rock or an opening under the eaves of a building. The nest is completed by early April and the bird then strikes up a song to attract a partner.

The female begins by decorating the nest with feathers and odd bits and pieces of vegetation and rubbish. Then she lays 4–6 light blue eggs at the rate of one every 24 hours. Incubation does not commence until the last egg is safely deposited in the nest. The duty is shared by both parents, the male during the day, the female at night. The fledglings hatch after two weeks and are fed by the adults with insects captured on the ground, in the foliage or on the wing. An English ornithologist has estimated that the parents visit the young about 300 times a day, in the course of which they provide some 1,200 food items. Considering the size of the starling population and the fact that the young are reared for three weeks, it only needs a rough arithmetical calculation to appreciate why the birds receive hospitality from farmers in the breeding season.

The fledglings continue to depend on their parents for some time, even after they begin to fly, but eventually form bands with others of their own age, by which time the adults are already busy with the next brood.

As the year goes on the starlings add fruit to their diet, making

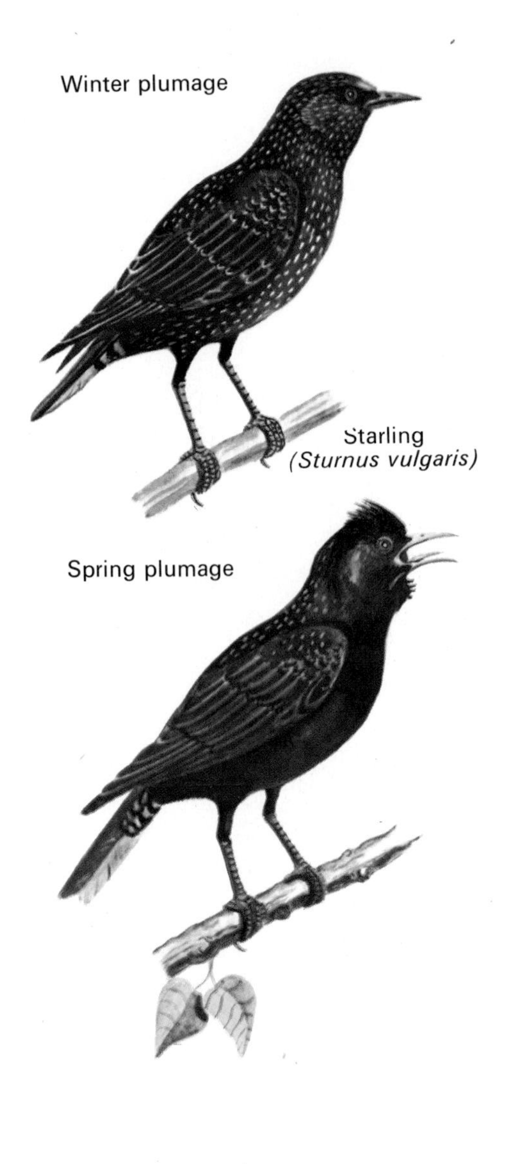

frequent raids on cherry orchards and later on vineyards. Some remain sedentary but others migrate in the autumn. The sedentary communities are those that live in western and south-western Europe. Those in the central and south-eastern parts of the continent are partially migratory. The true migrants are the starlings that nest in the east and north-east and abandon these regions after breeding.

The starling colonies that breed in Scandinavia, the Baltic countries, northern provinces of the Netherlands, northern Germany and northern Russia, head westwards in search of warmth. They settle in the southern Netherlands, northern France and the British Isles. The migrants from central and south-eastern Europe fly in a south-westerly direction and establish themselves in the coastal areas of the western Mediterranean. Most of the ringed individuals retrieved in Spain will have originated in Switzerland, Germany, Poland and Russia.

Before man disturbed their natural habitats, starlings wintering in the Mediterranean region fed mainly on the fruits of the arbutus and myrtle and on wild olives and figs. These species still feature prominently in the birds' winter diet and there seems to be a marked preference for red or black fruit. But several things have combined to modify these habits, notably the northern custom–dating from the Middle Ages–of providing the birds with artificial nesting sites, the deterioration of traditional habitats and the widespread cultivation of olives.

The olive growers declare war

In regions where olives are the main crop, the starling is an enemy to be destroyed at any cost. As evening shadows lengthen, an enormous black cloud of birds wheels up into the sky, turns and swoops down on the sedges and reeds lining a river or swamp and on the nearby thickets. Sometimes hundreds of thousands of starlings flock in these 'dormitories', providing local country-folk with an unrivalled opportunity of netting them in large numbers. Each morning they disperse, flying 40 miles or more in quest of food. By evening they are back home. Some ornithologists believe that this communal sleeping habit is a kind of safety device. The first individuals to take wing at dawn are invariably those that have fed plentifully the previous evening. They lead their less fortunate companions to the places where food is to be found in abundance and in this way all the members of the colony have a chance of satisfying their appetite.

Be that as it may, these huge flocks undeniably cause widespread damage in the Mediterranean lands. Tunisia has suffered more than any country from their invasions and there have been years when the greater part of the olive crop has been lost as a result of the birds' greedy activities. Gas and explosives have been used to combat them, but with scant success. Some idea of the scope of the problem may be gauged from estimates of flocks that are almost two miles in length, over 300 feet wide and 25 feet from top to bottom. In 1958-9 some 30 million starlings descended on the olive groves, costing growers the equivalent of seven million pounds sterling.

STARLING
(Sturnus vulgaris)

Class: Aves
Order: Passeriformes
Family: Sturnidae
Total length: $7\frac{1}{4}$–$8\frac{1}{4}$ inches (18–21 cm)
Wing-length: 5–$5\frac{1}{4}$ inches (12·3–13·4 cm)
Weight: 2–$3\frac{1}{4}$ oz (61–94 g)
Diet: fruit in winter, insects in summer
Number of eggs: 4–6, twice a year
Incubation: 15 days

Glossy black plumage in spring and summer, lavishly flecked with white in winter. The long, pointed beak is brownish-black in winter, lemon-yellow in summer.

The formation of a local defence association failed to improve the situation to any marked extent in the seasons that followed. Attempts at destroying the birds' roosting sites merely resulted in an increase of numbers in Sidi Saab, an area particularly noted for such invasions. Hundreds of local people were recruited in an effort to deny the starlings suitable landing places but it was a hopeless proposition. Eventually the government was forced to intervene, sending a team of dynamite experts to the region. Explosives were planted among the tamarisks where the birds were roosting and as soon as they settled down for the night the charges were set off simultaneously. Unfortunately the results were out of all proportion to the trouble and expense involved in this military operation. Out of tens of millions of roosting birds only some 350,000 were killed. Professional bird-catchers armed with nets could have done much better.

A more sophisticated method of fighting the winged invaders has recently been developed whereby tapes have been prepared on which the characteristic alarm cries of the species have been prerecorded. These have subsequently been run to coincide with the birds' attacks on the olive groves. Modest successes have been achieved by such means, but growers have realised that a more effective method is to reduce the number of births, either by stopping up cavities where nests are normally built or by allowing the birds to breed and later destroying the eggs.

One cannot disagree with the basic argument that the starling is a potentially destructive bird, but it is easy to understand why it should be so popular as an individual. It has one particularly diverting habit, namely the ability to imitate the song of other species with which it associates, as well as to mimic the natural sounds of its surroundings – as, for example, the splashing of a fountain. There is an amusing story of a football match in Ireland which was continually interrupted by a noisy starling perched on a tree beside the pitch. The bird imitated the sound of the referee's whistle with such startling accuracy that the players simply did not know which way to turn!

Although flocks of starlings are fairly harmless, even helpful, in the summer when they consume large quantities of insects, they are a plague in many countries during the autumn when their food consists in the main of fruit. The vineyards and olive groves of the Mediterranean region are particularly vulnerable to the concerted attacks of these black-winged invaders.

Breeding and wintering zones of the starling.

Rooks apparently pair for life but this does not prevent the male going through a ritual display every spring for the benefit of his mate. He bows before her, wings drooping and tail outspread, and makes characteristic cawing noises. He then offers her morsels of food.

The rook

The rook (*Corvus frugilegus*) is a migratory bird which spends the winter in the Mediterranean area because, like the starling, it has been able to adapt successfully to a modified habitat. Gregarious by nature, flocks of these birds descend on cultivated fields where, alongside crows, they methodically devour seeds, worms and insects.

The rook is easily identified by the grey patch of naked skin around the base of the long, pointed bill, and the feathers on the legs which look like baggy trousers.

The flocks return to their breeding grounds in February, each pair coming back to the same tree and even the precise nesting site they have occupied for years past. It would appear that two rooks form a partnership that lasts for life. Colonies usually comprise less than 500 individuals but up to 2,000 birds have been known to band together, sometimes building 60 nests in a single tree.

Breeding and wintering zones of the rook.

The first task of the breeding pair is to repair the nest. The male does most of the building, collects material and guards the nest in case a neighbour nips in while his back is turned. The female sometimes helps.

Before the birds couple the male pays court to his partner by displaying to her with wings drooping and his tail spread fan-wise. A courting rook looks magnificent despite his sombre plumage. In response the female dips her head slightly, flutters her wings and stiffens her tail. Finally the suitor offers her some food. The two birds then retire to the nest and the mutual display of esteem continues for a while before they actually mate.

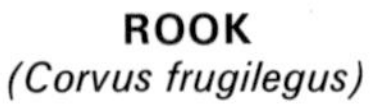

ROOK
(Corvus frugilegus)

Class: Aves
Order: Passeriformes
Family: Corvidae
Total length: $18\frac{1}{4}$ inches (46 cm)
Wing-length: 11–$13\frac{1}{2}$ inches (28–34 cm)
Weight- $\frac{3}{4}$–$1\frac{1}{2}$ lb (360–670 g)
Diet: seeds, fruit, insects, eggs, fledglings, small mammals, carrion
Number of eggs: 3–7
Incubation: 16–18 days

Black plumage, black feet, legs covered by thick ruffs of feathers. Long stout bill; greyish skin at base of bill, naked in adults.

Unlike many species that nest in colonies, not all the birds in a rookery lay their eggs simultaneously. Each female generally incubates 3–5 eggs (sometimes 6–7) for 16–18 days. Throughout this period and for the two weeks that the young are nestlings, the male feeds the entire family. The diet is extremely varied and changes in accordance with the locality and the time of year. It includes seeds, potatoes and other root vegetables, insects, worms, fruit, eggs, fledglings of other species, small mammals and carrion.

A study of the feeding habits of the species in the United Kingdom by the British ornithologist W. E. Collinge, based on the examination of the stomach contents of 1,306 birds, revealed that 59 per cent of digested food consisted of vegetable matter, the remaining 41 per cent being animal. But a separate analysis of food consumed at different times of the year has shown results very similar to those of starlings in their winter quarters.

When the fledglings are approximately two weeks old, the mother joins the father in looking for suitable food. After another fortnight has passed they are ready to be instructed in the rudiments of flying, being encouraged to bridge the gap between one branch and another with suitably placed titbits. Within a few days the youngsters are skilful enough to venture out into the fields with their parents but continue to be fed by them. After they are sufficiently experienced to leave the breeding ground (appropriately known as a rookery) it may continue to be used as a dormitory but is likely to be left deserted until the following spring. The young birds form flocks that disperse in all directions and when autumn comes the rook populations of eastern Europe set out on their migrations to the west and south-west. The western populations are, however, sedentary.

Although both belong to the same species and live in Europe, the carrion crow (*above and below, right*) and the hooded crow (*below, left*) differ in appearance and habit. The former is completely black and inhabits the western parts of the continent. The hooded crow has a black and grey plumage and is found mainly in eastern Europe. The two subspecies often interbreed.

The three species of pigeons found in the Holarctic region all have grey plumage with red and green reflections but may be identified by the presence or absence of white patches and black bands.

PIGEONS

Class: Aves
Order: Columbiformes
Family: Columbidae
Diet: mainly seeds but other vegetable matter taken
Number of eggs: 2
Incubation: 14–17 days

WOOD PIGEON
(Columba palumbus)

Total length: 16¼ inches (41 cm)
Wing-length: 9–10½ inches (22·5–26·2 cm)

Easily identified by white patches on sides of neck (absent in young) and white bar, prominent in flight, on either wing.

ROCK DOVE
(Columba livia)

Total length: 13 inches (33 cm)
Wing-length: 8–10 inches (20·3–25 cm)

Smaller than wood pigeon. Underside of wings white. Rump white. Two prominent black bars on each wing.

STOCK DOVE
(Columba oenas)

Total length: 13 inches (33 cm)
Wing-length: 7¾–9 inches (19·6–22·8 cm)

No white patches on body; faint black marks on wings.

Pigeons of town and country

Pigeons are familiar birds in both urban and rural areas and of the three European species the best known is the rock dove (*Columba livia*). This is the bird which has adapted so successfully to life in large cities, perching on public buildings and strutting through squares and parks, demanding food from visitors. Although tourists find them entertaining, they are not so popular with local residents and authorities because of their dirty habits. In the wild they congregate in dry localities, breeding among rocks and seldom settling on trees.

The wood pigeon (*Columba palumbus*), which has only been seen in numbers since the 18th century, is more frequently found on cultivated land. As its name suggests, it is a bird of the forests, roosting and nesting in trees.

The stock dove (*Columba oenas*) prefers the fringes of woods and open country with scattered tree cover, nesting in hollow trunks or rock crevices.

Of these three pigeon species only the first is sedentary. In the case of the other two, populations from southern and western Europe are sedentary whereas those living in eastern and northern countries fly south for the winter.

The melodious larks

All the birds commonly known as larks belong to the family Alaudidae but the many species vary considerably in appearance and behaviour. Five of the best known larks of the Holarctic region are the skylark (*Alauda arvensis*), the woodlark (*Lullula arborea*), the crested lark (*Galerida cristata*), the short-toed lark (*Calandrella cinerea*) and the calandra lark (*Melanocorypha calandra*). These birds are found on all types of open ground from tundra to desert and make up a high proportion of the bird population of steppe and farmland. The only exception to the rule is the woodlark which is more frequently found on the edges of woods and forests or on tree-covered hillsides, readily recognised by its mellow liquid song. The other four species, however, are essentially ground birds, laying their eggs in a

shallow depression which is sometimes lined with grass to make a proper nest.

Birds of the forest noisily proclaim their territorial rights from an elevated position in a tree. The larks are equally vocal, sometimes chirping from the top of a ridge or hillock but more typically lifting their voices in jubilant song as they soar vertically from the ground and plummet down to earth again. The song of the skylark (Shelley's 'blithe spirit') is unmistakable–a gushing torrent of trills and warbling notes, endlessly varied in rhythm and cadence and singularly beautiful.

This tendency to sing in mid-air, often at high altitude, is a characteristic adaptation to a treeless habitat. Its obvious disadvantage is that it attracts the keen attention of a multitude of winged predators. Were it not for these clear signals larks would be virtually immune from the attacks of birds of prey, for the cryptic coloration of their plumage–basically mottled brown or black on the back–provides excellent concealment as they crouch in the grass or corn.

The crested lark and calandra lark are sedentary species, whereas the woodlark and skylark (both of which breed in Britain) migrate to the south and west every autumn. They travel in formation both by day and night, remaining in flocks for the entire winter and feeding principally on vegetation. As they fly back to the breeding grounds the following spring each pair take possession of a separate piece of territory. In the breeding season and during the summer the birds' diet consists in the main of insects and other invertebrates.

The feral pigeons which are such familiar occupants of city squares and parks are rock doves. Inoffensive in the wild, they are so prolific that sheer weight of numbers makes them less popular in towns where they foul buildings and public monuments. No really effective method of limiting the population has yet been devised.

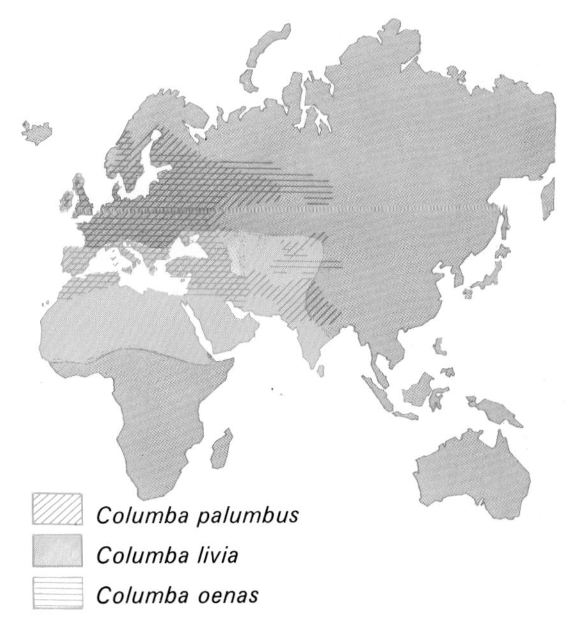

CHAPTER 6

The peregrine falcon: ace of bird hunters

Many birds pass the winter or remain to breed in the corn-growing flatlands of the Mediterranean, particularly in central Spain. There are no longer any large ungulates to be seen in these deforested regions and even the normally prolific rabbits have to struggle for survival here. But the birdlife in these parts is splendidly varied and interesting.

With the addition of the migratory species, there are birds in plenty all the year round. The resident communities include sand-grouse, stone curlews, great and little bustards, partridges, quails and calandra larks. The rock doves, originally accustomed to nesting in hills and mountains, have adapted with remarkable ease to this modified habitat, and tower-shaped dovecotes are typical features of the Castilian landscape. With man's assistance these pigeons have become very numerous and may eventually total more than all other birds in the area put together.

These sedentary species which nest and feed on the plains are reinforced at different seasons by migrants–spotless starlings in summer, turtle doves in spring and summer, common starlings, wood pigeons, lapwings and ducks in winter.

So many birds inevitably attract crowds of predators. Eggs, fledglings and adults alike are devoured by carnivorous mammals, reptiles and birds of prey; but one raptor deserves to be singled out as enemy extraordinary–the formidable peregrine falcon (*Falco peregrinus*).

It is true that the merlin also confines itself to birds but only to those smaller than quails and this raptor is only a winter visitor to these parts. Other typical birds of prey of the Mediterranean region–eagles, kites, hawks, buzzards and hobbies–attack birds, but merely as part of a much wider food range. The peregrine falcon is therefore the only all-season, exclusive hunter of birds.

Facing page : The peregrine falcon is a powerful bird of prey capable of killing species larger than itself, such as the little bustard. It eats as much as it can on the spot but never returns to finish off the remains of its meal.

Apart from the great bustard, which because of its size is virtually immune, every species is fair game for this, the most beautiful and rapid of all birds of prey. As a result of its specialised food habits the peregrine falcon has no direct competitors and flourishes here as in no other part of the world, with the possible exception of the Pacific islands where colonies of petrels and guillemots provide it with an inexhaustible supply of food.

The poison menace

Thanks to the rapidity of their flight and the efficacy of their hunting techniques peregrine falcons have formed colonies all over the world, with the exception of the Antarctic regions and a large part of South America. Wherever they can find suitable food, namely on relatively open ground, whether it be Arctic tundra, Asian steppe, American prairie or African desert, the birds will breed and prosper. In recent years, however, a grave threat to their future has arisen in the form of pesticides. The increasingly wide use of these poisonous substances has had harmful, sometimes disastrous, effects on many innocent forms of wildlife. The poisons work their way through all the links of a food chain and when the structure is complex – comprising a large number of consumers at different levels – the virulence of the toxic substances increases with each ascending stage. In its various habitats the peregrine falcon represents the top and final link in the chain and is for this reason especially at risk.

The birds on which the peregrine falcon preys feed in the main on seeds, insects or fishes. Since pesticides are normally used on crops in most areas, few birds, whatever their diet, can avoid a certain proportion of poison entering the body. Although the immediate effects may not be harmful the organism gradually accumulates larger and stronger concentrations of toxic substances which may have catastrophic long-term consequences (sterility being one of several symptoms). The poisons that coat seeds and which are present in the bodies of insects and fishes are automatically passed on into the bodies of birds in the normal course of feeding. It stands to reason, therefore, that the peregrine falcon, which preys on all these birds, absorbs the largest possible dose of accumulated poisons and that this may go far beyond the acceptable danger level.

This is why particularly careful watch is being kept on the falcon population of the Spanish plains. It is only in recent years that growers of cereals and leguminous crops on the Tierra de Campos, in La Mancha and in Estremadura have resorted to pesticides. The rock doves which are the raptor's favourite prey in these parts have not so far shown any obvious ill effects but it is likely that increased amounts of poison will eventually cause damage to them. A more immediate threat may come from the starlings and thrushes that originate in olive-growing regions where DDT is extensively used. These birds are captured and eaten from time to time by peregrine falcons and the fear is that there may be a delayed reaction, similar to that which occurred with falcons in North America and the British Isles, where in certain district numbers were reduced by between 60 and 90

The peregrine falcon is the most popular of trained birds for falconry. It is often used to hunt birds which frequent airfields and which pose a threat to aircraft and passengers by being sucked up into jet engines.

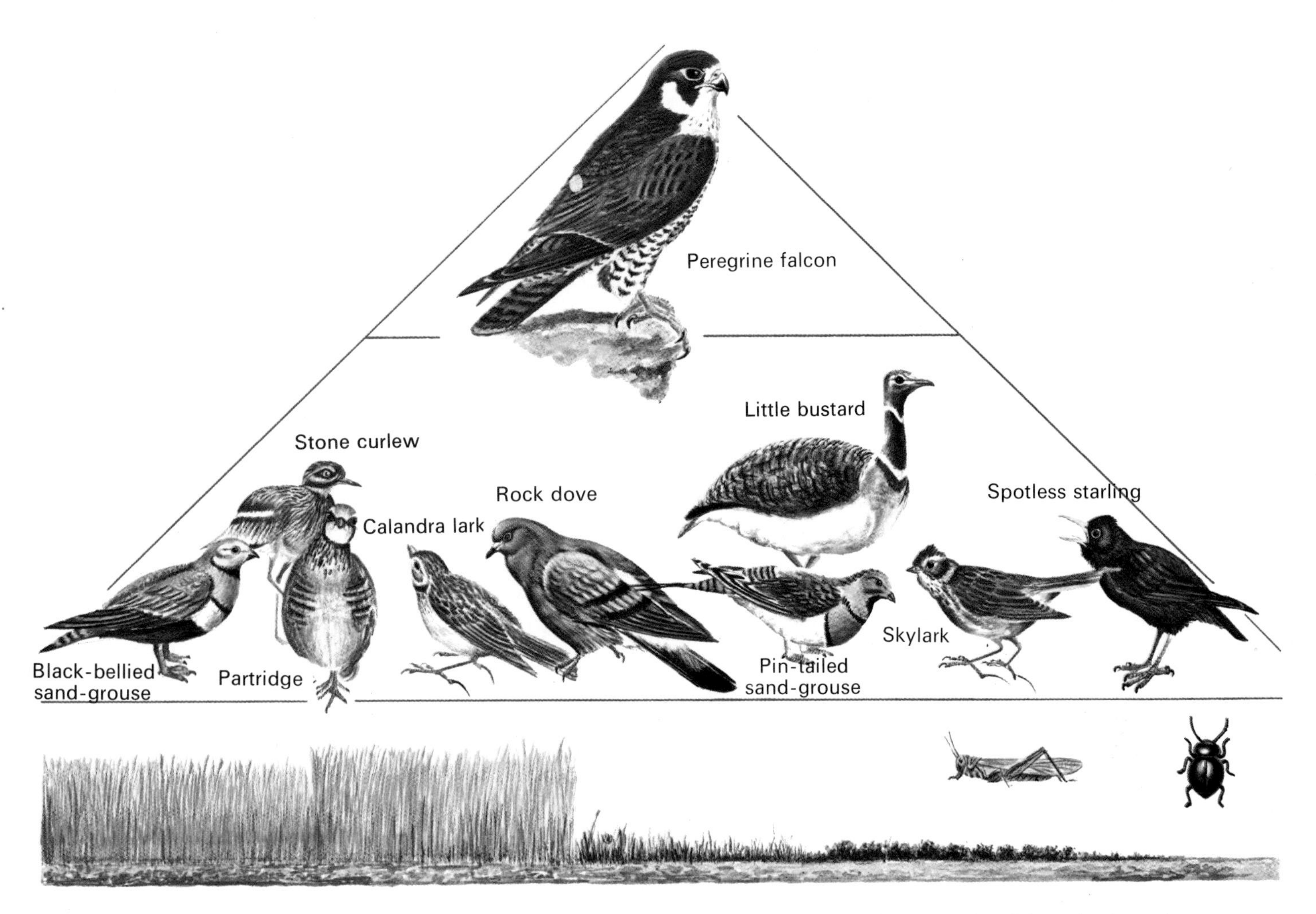

The peregrine falcon is the only raptor which feeds exclusively on other birds and therefore occupies a position at the apex of its ecological pyramid, as in the grasslands and cornfields of the Mediterranean region. This apparent advantage has nevertheless resulted in the gradual disappearance of the species in most parts of the world. Pesticides used on crops are absorbed by seeds and insects and then find their way into the bodies of granivores and insectivores—species regularly hunted by the peregrine falcon. The latter therefore accumulates a massive amount of poison in its body which may ultimately cause sterility.

per cent—an alarming trend. Disturbing signs are already evident. For the past twenty years records have been kept in Spain of a control group of fifty pairs of falcons. Recent statistics show that the birds breed less frequently, that they are laying fewer eggs and that there are diminishing numbers of survivors from each brood. These are precisely the warning signs that were received some years ago from the United States at a time when the peregrine falcon population in that country was still quite large. Today that population has almost vanished. A repetition of this sequence of events in Spain would be catastrophic for the species as a whole, for Castile is regarded by many ornithologists as one of the last seemingly secure refuges of this magnificent bird of prey.

Beauty and power

For thousands of years the peregrine falcon has been esteemed for its hunting prowess and prized by falconers for its singular combination of strength and beauty. With its solid, round head, short neck and compact body, the bird is a natural marvel of aerodynamic design. The long wings are pointed (almost triangular in form) and when folded reach the tip of the tail, which is short in comparison with that of other falcons. The broad-shouldered body is supported by feet that are larger and more powerful than those of most raptors. The front toes are long and sturdy with sharp claws, the lower parts of the toes being furnished with rounded knobs, facilitating the capture of prey. The hind toes are shorter and the claws less developed. The tarsi are naked and thickset, with prominent joints.

Seen from the front, the peregrine falcon's head is heavy, almost square. The broad back and wide shoulders are in strong contrast to the narrow waist. As it grows older the bird tends to put on a little weight and the feathers become progressively shorter after each moult. Both these phenomena are an advantage to a raptor which relies on powered flight rather than gliding; and age of course brings not only improved performance but also added experience.

The dominant colours in the young falcon are brown and reddish-brown. The back is uniformly dark brown although sometimes light brown at the edges. Two reddish marks extend down the neck at an angle on either side of the head. The moustachial stripes vary in size and colour. In some cases they are broad and black, covering the cheeks and forming a band round the throat; in others they are just two thin chestnut lines. Dark tear-like marks streak the chest, belly and flanks, the ground colour being reddish or ochre. Cere and brows are blue-grey during the first few months; the feet and tarsi are usually greenish-yellow.

This plumage undergoes a complete alteration when the bird is about one year old and takes on the predominantly slate-grey hues of the adult. There is not a great range of individual variation. The back is a fairly uniform blue-grey, the head and neck slate-grey. The face is very distinctive, framed as it is by the two black moustachial stripes which extend down to the throat behind the beak on either side of the head, setting off the white bib-like area of the cheeks, throat and upper part of the chest. The bill is short, conical and very broad at the base, with a prominent notch on each side of the upper mandible. The original yellow colour turns to orange as the bird gets older and the tip is blue-black. The nostrils are circular, flared and furnished with a characteristic tubercle. The eyes are very dark because the deep chestnut iris is indistinguishable from the pupil, but they are set off by a surrounding bare patch of bright yellow skin. There is no glint of cruelty in them; in fact they convey an impression of untroubled serenity.

The lower parts of the body are much lighter in colour, ranging from buff to pink. The belly is boldly striped (no longer streaked as in the immature bird) with horizontal bluish lines. The flanks and rump, equally prominently striped, are greyer and this pattern continues to the tail which has a pale border. The feet and tarsi are yellow.

A world-wide range

The peregrine falcon, combining strength, stamina and speed to perfection, has utilised these qualities to make a home in almost every part of the world. The species has established itself not only on the mainland of all five continents but with its ability to cover immense distances across seas and oceans has also settled in the Cape Verde and Falkland Islands in the Atlantic, and the Volcano and Solomon Islands in the Pacific.

This does not mean that the bird will breed wherever it has

Facing page : During its first year of life the young peregrine falcon (*above*) has a predominantly brown or reddish-brown plumage, which later turns slate-grey with prominent black bars. The raptor has large, dark brown eyes, rounded nostrils and a strong, short bill with a toothed projection on either side.

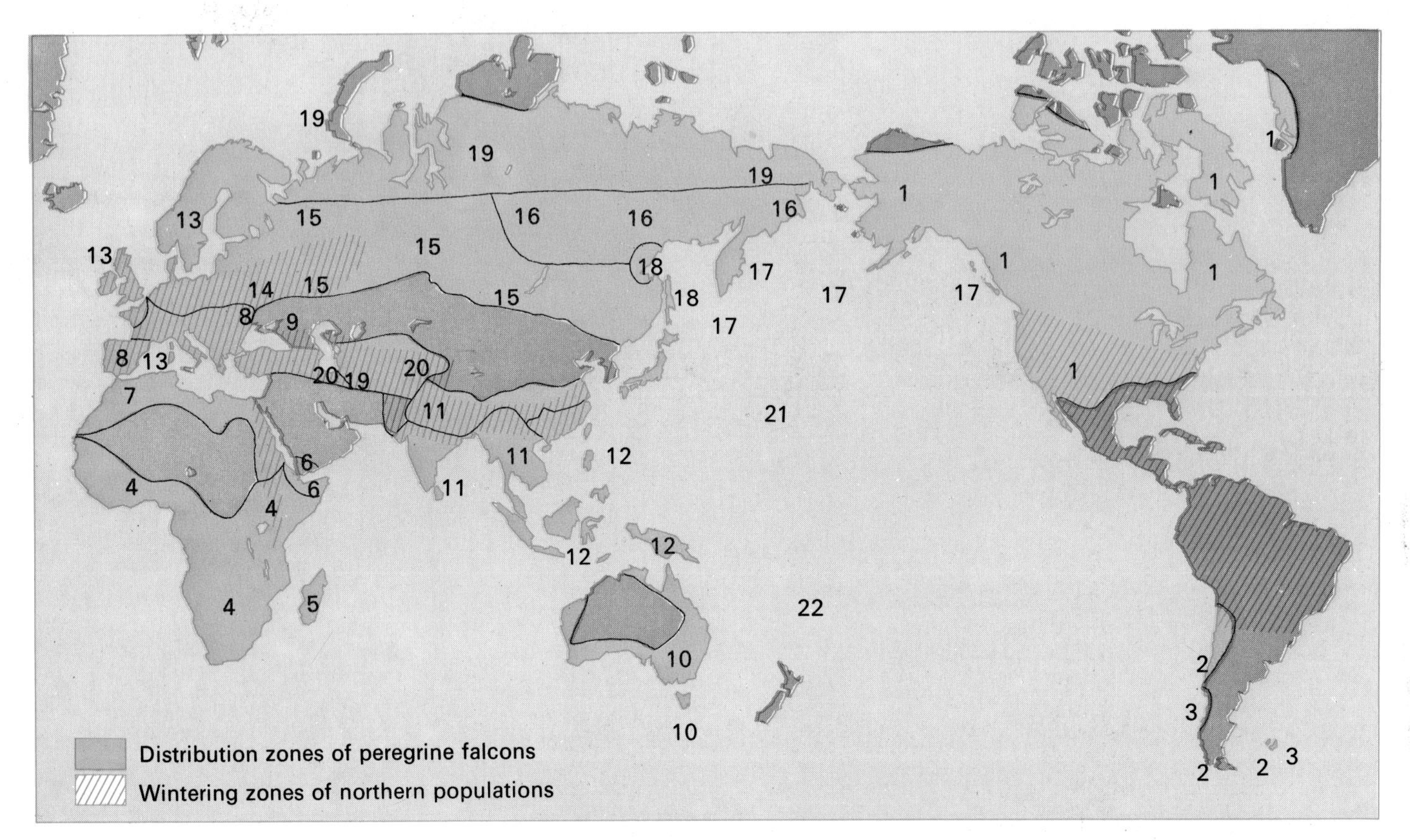

The Russian ornithologist Dementiev has identified twenty-two subspecies of peregrine falcons (*Falco peregrinus*). They are found all over the world, except in the Antarctic, at the North Pole and in parts of South America. The numerals indicate the wintering zones of the following subspecies: 1) *anatum*, 2) *cassini*, 3) *kreyenborgi*, 4) *perconfusus*, 5) *radama*, 6) *arabicus*, 7) *pelegrinoides*, 8) *brookei*, 9) *caucasicus*, 10) *macropus*, 11) *peregrinator*, 12) *ernesti*, 13) *peregrinus*, 14) *germanicus*, 15) *brevirostris*, 16) *kleinschmidti*, 17) *pealei*, 18) *pleskei*, 19) *calidus*, 20) *babylonicus*, 21) *fruitii*, 22) *nesiotes*.

been sighted. It will only reproduce in temperate zones where there is sufficient humidity and where there are cliffs and rocks. According to K. H. Voous the ideal temperature required by a pair of peregrine falcons for breeding ranges from a minimum of 5°C (41°F) to a maximum of 32°C (88°F). At other times of year they can endure greater extremes of cold and heat. In the Arctic, arriving when the tundra is not yet completely frost-free, they withstand the icy north wind that sweeps across the barren wastes at night; and by contrast, the birds of a semi-desert subspecies in Sudan spend the hottest hours of the day sheltering in the shade.

Peregrines are also found at varying altitudes, from cliffs along the sea-coast to peaks of 13,000 feet or more, as in the Himalayas. It is comparatively rare, however, for the raptors to roost at such great heights and the largest populations prefer to remain at sea level or on ground which rises to 3,000 feet or thereabouts.

The fact that peregrine falcons are to be found at latitudes as widely divergent as the Alaskan tundra and the shores of Tierra del Fuego and in regions where temperatures vary from extreme Arctic cold to stifling tropical heat, indicates that some, though by no means all representatives of the species, are migratory, notably those subspecies that are subjected to the most arduous, rigorous climatic conditions. In fact the communities inhabiting the Eurasian steppes normally leave these regions during the spring and summer. By September they have arrived in their distant winter quarters in India and East Africa. Individuals from Eurasia have been identified in the neighbourhood of Lake Naivasha in Kenya and in Tanzania's Ngorongoro Crater. As for the peregrine falcons of Alaska and Canada, they fly south for the winter to Florida and South America, even beyond the equator.

Yet even peregrine falcons nesting in warmer climes, as in the Mediterranean region, migrate in certain circumstances, when, for example, their nesting sites are high above sea level. Thus in the Pyrenees and other mountain districts of Spain and Portugal, the birds abandon their eyries in September and do not return until the following March.

It is interesting to note that there is a marked physical difference between the peregrine falcons that migrate and those that are sedentary. The individuals that are forced to undertake arduous journeys have a slimmer, more streamlined body and somewhat longer wings and tail. Thus they are better equipped for gliding flight, able to take advantage of the wind to travel considerable distances. Speed is not as important to them as stamina. In fact estimates of the speed attained by falcons journeying from continent to continent tend to be greatly exaggerated. In normal weather conditions the average speed is probably not more than a steady 30–40 miles per hour.

The sedentary peregrine falcons tend to be more square in shape and rather more robust. Their shorter, narrower wing and tail feathers are designed to provide the necessary speed and manoeuvrability for catching smaller prey on the wing.

Since the peregrine falcon population is widely distributed throughout the world, it is hardly surprising that there should be a number of different subspecies, characterised mainly by variations in colour, shape and size. Ornithologists disagree as to how many such races exist. C. Vaurie, for example, has identified twelve, whereas G. P. Dementiev distinguishes twenty-two (see chart on facing page). It is a difficult and controversial subject. Thus the Tierra del Fuego peregrine falcon (*Falco peregrinus kreyenborgi*), which was previously considered to be a subspecies, is nowadays acknowledged by an increasing number of ornithologists to be a species in its own right, following detailed study of museum specimens. It is also thought that the Barbary peregrine falcon (*Falco peregrinus pelegrinoides*), which has adapted to the climatic extremes of the North African desert and which was also formerly classified as a subspecies, is likewise a species proper. Nevertheless, although a great deal of research has been done on the problem, no conclusive information has so far been published covering the reproductive behaviour of these birds, this being the necessary criterion to establish whether or not they are entitled to specific classification.

It seems to have been proved that the individuals that nest in Spain and Portugal belong to the subspecies *Falco peregrinus brookei*–the bird often referred to as the Mediterranean peregrine falcon. It is possible, however, that representatives of the Barbary race sometimes nest as well in the south-eastern corner of the Iberian peninsula.

In September the sedentary population in Spain and Portugal is in any event joined by migratory subspecies from the north even if the latter only regard the region as a stepping stone to winter quarters in Africa. Among these regular autumn and spring visitors are *Falco peregrinus germanicus*, an inhabitant of central Europe, and *Falco peregrinus peregrinus*, which breeds in northern Europe and is often seen in the north-western

The different subspecies of peregrine falcon show variations in size and colour of plumage. The migratory Siberian peregrine falcon (1) is stronger, slimmer and paler in colour than the sedentary representatives of the Mediterranean subspecies (2). The Barbary peregrine falcon (3) has a large head, short tail, small wings and pinkish-brown plumage.

Facing page: Although it breeds on rocky terrain the peregrine falcon hunts over open ground, unbroken by hills or cliffs. In order to survey the surroundings and keep a lookout for prey the raptor sometimes settles on a branch of a dead tree or on a sand dune.

tip of the peninsula. The great migrants of the Asiatic tundras, the Siberian peregrine falcons (*Falco peregrinus calidus*), rarely pay visits to Spain for their seasonal travels take them directly across the eastern Mediterranean and their winter quarters are in southern Asia and East Africa.

Ancient falconers used to give distinctive local names to the various falcon subspecies which they trained for the hunt, based principally, so it would seem, on colour differences. It is noticeable that the peregrine falcons of central and northern Europe possess darker backs than those of the Mediterranean region and that the latter have pinkish markings on the neck. The backs of the North African birds are browner and the underparts contain a greater proportion of red. The rare Siberian subspecies tends to be pale grey all over and is characterised by a very narrow moustachial stripe.

Rock fortresses

Despite threats to their future there are still enough adult peregrine falcons in central Spain to permit ornithologists to make detailed surveys of territorial and reproductive behaviour. As a rule these raptors prefer to choose nesting sites along rocky river valleys, each eyrie being occupied by a pair of falcons which seldom journey more than a few miles in any direction in quest of food. But the size of the hunting territory will vary according to the lie of the land. Rivers flanked by steep rock walls are not characteristic features of the Spanish landscape, so that a pair of falcons may have to roam quite an extensive tract of territory in order to find an ideal nesting site. The birds seldom settle in trees and equally rarely lay their eggs on bare soil, although on two occasions observers have found traces of an eyrie on a steep mountain slope where the tracks of sheep have formed slight ridges on the ground. Typically the nest is on a rock ledge, usually high up but it may be near the ground. At times the old nests of raven, crow, jackdaw or herring gull are used.

The density of the species on the vast corn-growing areas where there is a rich avifauna, is surprising in view of the fact that adult peregrines frequent especially rocky prominences which are here little in evidence. The isolated rocky nesting sites normally chosen by the raptors are few and far between in these cultivated areas. In fact the falcons which invade the plains from August onwards are for the most part young birds expelled from the eyrie by their parents after completing their hunting apprenticeship.

By the middle of September these wandering youngsters begin to be joined by groups of migrating subspecies flying down for the winter from northern and central Europe. The numbers of migrating falcons are nowadays comparatively small, for the extensive use of DDT and other pesticides in the areas where they breed has recently taken a terrible toll. Now the immature sedentary falcons and the young migrants from abroad spend the night together on the open plains, choosing some such vantage point as a heap of stones, a hillock or even an old tree. Meanwhile the sedentary adults seldom venture far from the eyrie and do not

PEREGRINE FALCON
(Falco peregrinus)

Class: Aves
Order: Falconiformes
Family: Falconidae
Length: $15\frac{1}{4}$–20 inches (38–50 cm)
Wing-length: male $11\frac{1}{2}$–13 inches (29·3–32·2 cm)
female $13\frac{1}{2}$–$14\frac{1}{2}$ inches (34·4–37 cm)
Wingspan: $32\frac{1}{2}$–$44\frac{1}{2}$ inches (83–113 cm)
Weight: male $1\frac{1}{4}$–$1\frac{1}{2}$ lb (570–650 g)
female 2 lb (850–1,000 g)
Diet: exclusively birds
Number of eggs: 2–6, normally 3–4
Incubation: 28–30 days

Adults
Individuals belonging to the typical subspecies *Falco peregrinus peregrinus* have a blue-grey back, sides of head black, large black moustachial stripes, white throat and pale chest with black bars. The wings are bordered with black. The tail, similar in colour to the back, has brown stripes. Feet and cere are yellow, iris dark brown. The female is larger than the male.

Young
Initially covered with whitish down, then chestnut plumage on upper parts, lighter on chest and belly with dark brown streaks. Slate-grey cere, grey or greenish-yellow feet; dark brown iris.

mingle with the winter visitors. Male and female pair for life and never yield their eyrie to intruders. They return every evening to sleep and sometimes visit the eyrie during the day for brief periods of rest in order to recover their strength for another hunting foray. These local communities have lived in the same area for as long as anyone can remember and certain huge rocks have been occupied by one family from generation to generation. Experience has familiarised the birds with every corner of their habitat so that they enjoy virtually complete security. In the event of one of the adult birds dying or being killed, particularly if the victim is the female, its place is promptly taken by another. Evidently the male defends the eyrie against all rivals and soon manages to find a new partner. Felix Rodriguez de la Fuente has watched peregrine falcons guarding their nesting site and has seen solitary males driving away adults of the same sex whilst tolerating the exploratory visits of unattached females.

The monogamous habits of the species ensure that territories—or at least areas reserved for nesting and breeding—do not overlap. Felix Rodriguez de la Fuente points out that families, when sleeping or resting, remain a mile or so apart from each other, but admits that it is difficult to ascertain whether or not the birds infringe on one another's hunting grounds. As an expert in the sport of falconry he has taken trained birds into districts occupied by wild peregrine falcons and has seen the latter launch determined attacks on the intruders, clearly with the view of expelling them from private territory. One of his young female birds was violently attacked by a wild falcon, even though the latter's eyrie was about a mile away. The claws of the aggressor ripped a deep wound in the trained bird's back, damaging a lung.

There are few places left outside Spain where the peregrine falcon today enjoys such security or where it is found in such numbers. The only exceptions are some of the rocky islands (notably Queen Charlotte Island) off the coast of British Columbia where the abundance of sea birds and suitable nesting sites provide ideal conditions for the subspecies *Falco peregrinus pealei*, the density of the population being about twenty pairs to the square mile. Here, as elsewhere, the extent of a piece of hunting ground is directly related to the amount of prey available and the nature of the terrain for breeding purposes. Thus in the river valleys of Alaska eyries may be two or three miles apart, whereas in tropical Africa the distance between two sites will be nearer to ten miles. It is hard to produce accurate figures. All one can say is that the area of a peregrine falcon's territory generally ranges from two to five square miles: but in the British Isles it is more likely to fluctuate between ten and fifteen square miles.

We have spoken of the nesting habits of the Spanish subspecies, which are typical of many other peregrine falcons of temperate zones. But behaviour is much influenced by climate and other factors. In sparsely populated regions, for example, where the raptors are virtually undisturbed by man, they have so far adapted to local conditions as to lay their eggs directly onto the ground, usually on top of a small mound or on a slight slope so that the female can keep a good watch on the surroundings. In Alaska, Canada, Finland and on the tundras of northern Asia, many

peregrine falcons rear their fledglings at ground level. What is more unusual, in northern Germany, Russia and Siberia, sedentary members of subspecies such as *Falco peregrinus germanicus* even nest in trees. Ornithologists in the Baltic region once counted 45 eyries placed in the normal fashion on rock ledges and 125 in the branches of trees.

As a rule nuptial territory is fairly restricted in area and the nesting site is usually found in the centre of this zone.

The peregrine falcon never builds its own nest. Favourite nesting sites are flat rock ledges or crevices well off the beaten track.

Courtship and reproduction

The peregrine falcons of the Iberian peninsula embark on their spectacular nuptial displays in February. During the winter male and female will have occupied the same eyrie as in previous seasons but their relationship has been marked by courtesy rather than affection. The two birds occasionally exchange a peck of recognition when one of them returns to the eyrie after hunting or when their paths happen to meet in the course of such expeditions, but otherwise they are not especially demonstrative. But when the days begin to draw out, even though frost may still cover the ground in the morning and the night temperature is still hovering around the zero mark, both birds start preparing themselves for the complex round of activities which will reach a climax some weeks later. They begin by making a series of short, cautious flights close to the selected nesting site, soaring upwards in broad circular paths, then diving down again at breakneck speed. The male often hunts on behalf of his partner and feeds her with captured birds, either in the eyrie itself or in mid-air.

On those occasions when the male flies back to the nesting site with a food offering, the female – as happens in many other species – adopts typically juvenile postures, swelling her feathers, spreading her wings and cheeping faintly in a manner characteristic of a newly born fledgling. This food-receiving ritual frequently takes place on the bare rock ledge where the eggs are soon to be deposited. Felix Rodriguez de la Fuente has noted that it is the

Facing page : The compact body, broad chest, pointed wings and square tail of the peregrine falcon help to make the bird a master of the skies, long prized by falconers for its hunting prowess.

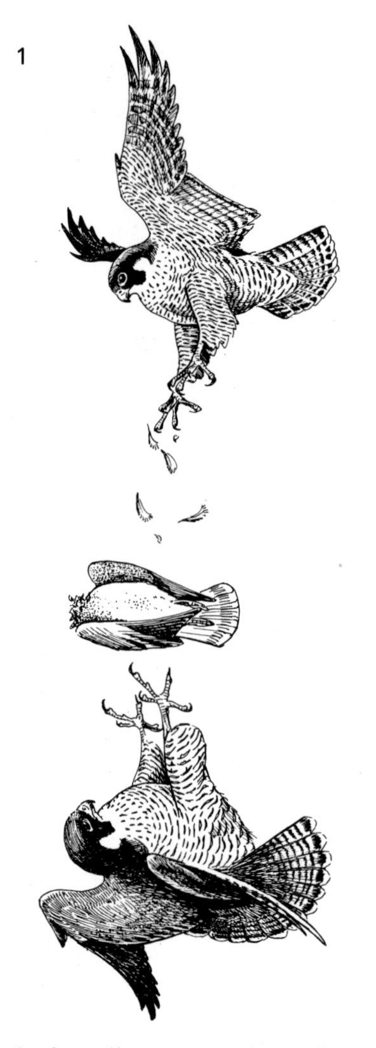

During the breeding season the male peregrine falcon punctuates his courtship display by killing birds that are not normally hunted and offering them to his mate. Sometimes the female meets him in mid-air (1), turning on her back to receive the prey. Alternatively, the male brings the offering back to the ledge to be used as the nesting site (2), the female reverting to juvenile mannerisms, complete with chirping.

Facing page : The abandoned nest of a crow or an eagle will serve very well as an eyrie for a roving peregrine falcon. While the male scours the plains for food, the sturdier female defends the eyrie and rips up the prey for her downy fledglings.

male who generally selects the site and who entices his partner to join him by tendering scraps of food and by adopting characteristic attitudes that are very similar to the movements made by the female during incubation.

The original choice of ledge or cavity for egg-laying is of primary importance for the survival of the brood in view of the fact that the adults never construct a proper nest for the protection of their young, being content merely to scoop away a layer of soft earth (should this be present) with their feet. The sheer rock faces on which they settle often seem to have no visible projections or inlets suitable for nesting and yet somehow the raptors manage to find a convenient platform for the purpose. But very often they will scour the area for a ready-made nest–usually one that has been built and subsequently abandoned by a member of the crow family–and this may be utilised for a number of years, to be relinquished only when the elements have battered it beyond repair.

In Spain peregrine falcons are normally found on cliffs, in deep river gorges, in the mountains of the central provinces, and on the steep hillsides overlooking the arid plains of Old and New Castile. Although it is rare, peregrine falcons' eyries have been discovered among the ruins of an ancient castle or in a demolished building. Elsewhere in Europe the raptors choose the customary rocky ledges and crevices but also appropriate the empty nests of crows, rooks, ravens, jackdaws, eagles and herons. In no cases have they been known to collect material for building their own nest.

Courtship begins before the birds settle into the nest. When first searching for a convenient nesting site, the male persists with his courtship ritual. From time to time the two birds perch on a rock to mate, showing hostility to others of their species.

In central Spain the female lays three or four eggs around the middle of March. Nests have been found containing two eggs but never five or six (as is the case with the subspecies inhabiting the eastern seaboard of the United States). The mother no longer strays from the eyrie and sometimes begins to incubate the eggs as they are laid so that the fledglings eventually hatch at intervals. The average duration of incubation is 28–30 days. In Spain it is usual for all four, or at any rate three, of the newborn falcons to survive. In Alaska the infant mortality rate is roughly 15 per cent. Recent reports, however, from the British Isles and other parts of Europe, as well as from North America, indicate a steadily rising death toll as a result of toxic pesticides.

The representatives of the subspecies *Falco peregrinus brookei*

Even when the fledglings' feathers begin to appear there are still traces of fluffy white down. But by now they are able to fly and be initiated into hunting methods. Several months later they will leave the eyrie for ever.

feed their offspring in the eyrie for about five or six weeks. During this time there is a clear division of labour between the parents, the male being responsible for finding food for his family, sometimes bringing back as many as eight items of prey to the eyrie, the female mounting guard over the rock or nest, flying out to meet her mate and uttering sharp warning cries if she sights an enemy in his absence. Female peregrine falcons have been seen to launch determined attacks on crows, eagles, foxes and even dogs that dare to roam too close to the family refuge. In regions where humans are seldom encountered the occasional passer-by risks being subjected to sudden attack as the falcon swoops down, checking its fall only a few feet above the head of the intruder.

In addition to defending the eyrie the mother plucks the feathers from the birds killed by her mate and then meticulously tears the bodies into manageable pieces for her offspring. In the event of the female's death the solitary male is apparently unable to combine hunting, food preparation and defence so that survival of the brood is problematical.

As long as the fledglings are still enveloped in down the mother continues to feed them by placing small pieces of meat in their mouths. Thanks to the equal distribution of responsibilities they are never left alone and always provided with sufficient food. While one parent attends to their needs, the other flies off on another hunting expedition for even though food is comparatively easy to come by at this time of year he must keep busy in order to satisfy the fledglings' voracious appetites.

It is probable that this division of labour in the course of rearing the young has determined the difference in size between the sexes. Thus a male weighing approximately 1¼ lb may mate with a female weighing 2 lb or more. The comparative lightness of the male is undoubtedly an advantage for hunting birds on the wing, whereas the heavier and more powerful female is better equipped to defend the eyrie against intruders such as crows which often attempt to filch the eggs and kill the fledglings.

The marked difference in size and weight between male and female is also of value for the species since the dual qualities of agility and strength complement each other so successfully that a very wide range of prey can be coped with, from tiny fledglings to birds weighing several pounds, including many that are several times larger than themselves.

The young falcons attempt their first flight at the age of five or six weeks but remain within the family circle and sleep in the eyrie for about another month. They continue to depend on the mother for protection and both parents gradually start instructing them in techniques of capturing prey. They start by bringing home dead birds and tossing the bodies about so that the youngsters can practise catching them in mid-air; later they get accustomed to venturing out alone in quest of living prey.

Once this period of apprenticeship is over the parents encourage their offspring to become increasingly independent so that they can prepare themselves, if necessary, for long and difficult journeys across unfamiliar terrain.

During the first year, while they still retain their covering of down, the young falcons do not mate, but in the course of the second year the adult plumage appears and they are then able to procreate. Because this sexual activity begins so early in life and since the species is comparatively long-lived (recent surveys have identified several twelve-year-old females and one aged eighteen years which were still capable of breeding) the peregrine falcon population–provided there are no outside adverse causes–is able to maintain its numbers fairly well and to keep pace with infant mortality losses. This balance of births and deaths is of course important for the future of a species which increasingly finds itself on the knife edge of survival.

Another natural advantage is the ability of the female peregrine falcon to lay another clutch of eggs in the event of the first being stolen or broken as a consequence of rolling off a narrow or sloping rock ledge. Many birds that nest among cliffs and rocks face this latter risk but the majority of sea birds cope with the problem by constructing a nest which affords adequate protection both to eggs and fledglings. Peregrine falcons, as we have seen, simply lay their eggs on the stone surface so that accidents of this nature are not uncommon.

Among adult peregrine falcons the mortality rate ranges from 15–30 per cent in most regions but in Spain the figure is–for the time being–comparatively low.

The two hundred mile-per-hour dive

The peregrine falcon is the nose-dive champion of the bird world, falling on its prey either vertically or at an oblique angle in a concentrated attack which strains its limbs and natural faculties to the limit.

It is immediately obvious that the shape, structure and functions of the raptor's body are in perfect harmony with its special

Pigeons form the basis of the peregrine falcon's diet although the proportion killed in relation to other birds varies according to season and locality.

method of hunting. The wings are long and slender with strong, resilient feathers, the second of which is larger and more prominent than the rest. Since the bird seldom needs to attempt aerial manoeuvres at slow speed the tail is relatively short, with small and compact feathers on the first finger.

The falcon normally gains height by allowing itself to be borne aloft on warm air currents, sometimes achieving added thrust by beating its wings in sharp and rapid rhythm, rather in the manner of a duck. Having risen to an altitude of 1,500–2,500 feet it looks down on its hunting grounds, ready to launch an attack as soon as suitable prey is sighted. But such an onslaught will not be indiscriminately directed against the first bird to loom into view, for like most raptors the peregrine falcon carefully selects its victim. Thus when a flock of pigeons flies far below the falcon will, circumstances permitting, dive on a young bird or one that is having some difficulty in flying, perhaps as a result of having lost some feathers. In spring the glowing nuptial plumage of the males of many species catches the eye not only of the intended partner but also of the hovering enemy. By summer the fledglings have hatched and it is unnecessary for the raptor to waste skill and energy on the adults; and when winter comes attacks will be concentrated on birds that are too old or sick to move very fast.

The ability of a peregrine falcon to spot and isolate individuals showing the slightest symptom of fatigue or weakness is quite astonishing. Experiments have been conducted whereby groups of starlings and pigeons have been released, among which are a few birds with a couple of wing feathers removed. On almost every occasion the falcon would swoop unerringly on such handicapped birds.

The hunting techniques of the peregrine falcon are to a large measure determined by the direction of the wind. With the wind behind it the raptor can more successfully control its dive, usually swooping down on its prey in an almost uninterrupted straight line. In this type of attack, launched at high speed, the impact of the falcon's feet is often sufficient to break the victim's wing or neck. The precision of the dive, appreciated by every falconer, is unbelievable, considering the velocity at which the bird is travelling as it strikes home. This may in fact be around 200 miles per hour, some authors claiming that under ideal conditions it may be measured at nearer 280 miles per hour. Of course the average speed at which the falcon flies is very different. When migrating, for example, the bird travels at about 30 miles per hour, which may be considered normal cruising speed. Although the raptor beats its wings to increase velocity when diving, it will alternate standard wing-beating flight with long, leisurely gliding actions when simply moving from place to place.

When the falcon flies against the wind it does not generally aim to kill the prey in mid-air but seizes it with its claws, usually taking hold at the bases of the wings and bringing it down to earth as quickly as possible, there to finish it off. The victim will probably be stunned or injured and therefore incapable of offering much resistance as the raptor deals the death blow with its bill, striking either at the neck or skull. The power of this blow normally causes instantaneous death, even in the case of such a

Facing page : In the Mediterranean region both wild and domesticated pigeons are killed in large numbers by a flourishing population of peregrine falcons. The primarily seed-eating flocks of pigeons cause considerable damage to crops so that these predatory activities are of value to farmers.

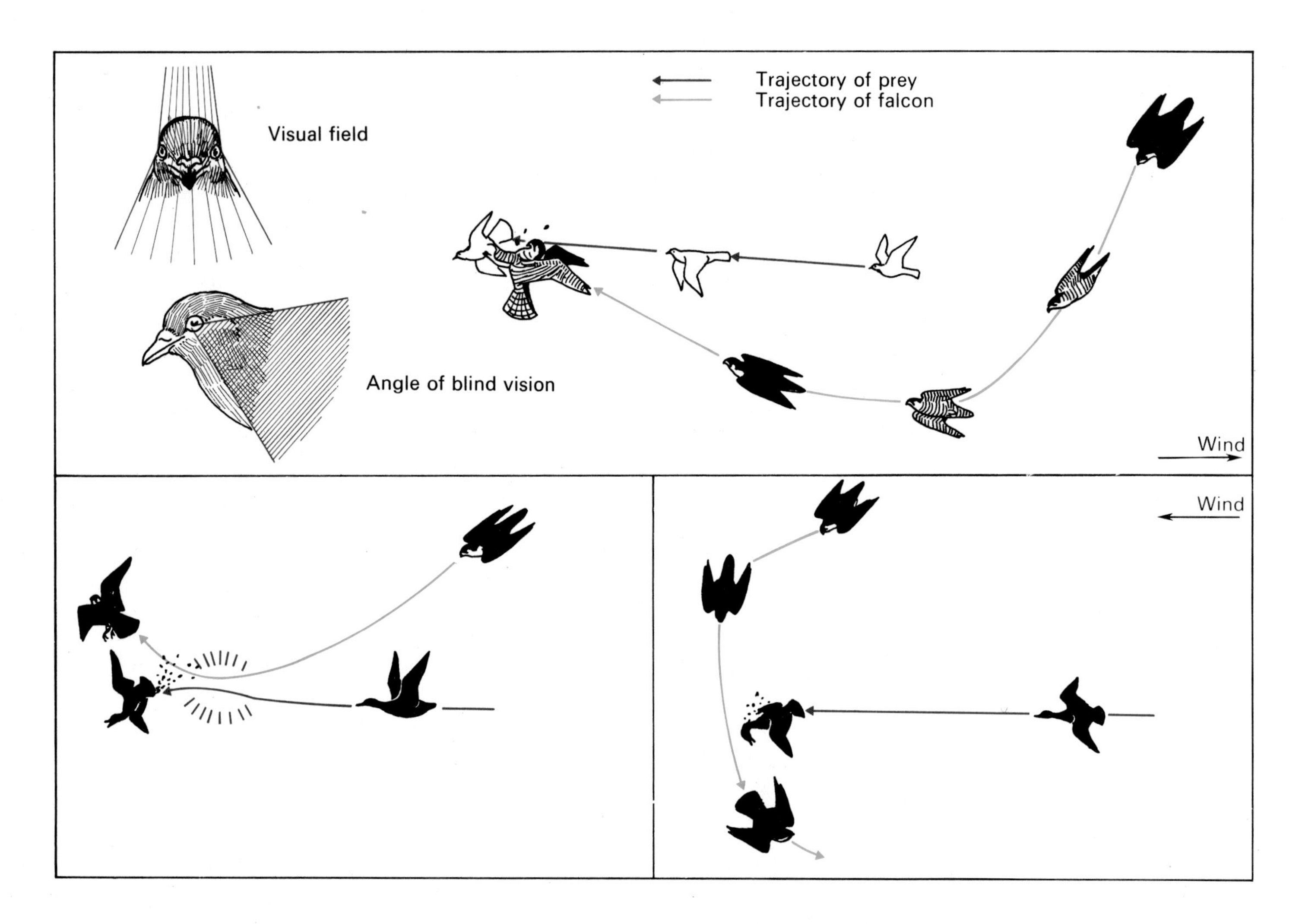

The peregrine falcon specialises in the capture of birds in mid-air and employs a variety of hunting techniques depending, to some extent, on the direction of the wind. When flying into the wind, as in the upper diagram, it will dive at an angle on its prey (in this case a pigeon) but check its fall at a point some distance behind and below the victim's flight path. It then tucks its wings close to the body and soars upwards to seize the pigeon from underneath, its line of attack so selected as to be invisible to the victim (the latter's field of vision extending forwards, upwards and sideways, but not to the rear). The lower diagrams show alternative forms of attack with a following wind, the victim in both cases being a duck. The falcon may either dive directly at the prey at a speed of around 200 miles per hour or swoop down and position itself ahead of the victim's flight path and provoke a collision. Either way the attack must be launched with deadly accuracy, so that the impact is sufficient to kill or injure the prey but not so powerful as to damage the raptor's own feet or wings.

heavy bird as the little bustard.

The peregrine falcon does not necessarily always dive in a straight line at its victim. Surprise is often as vital as speed. If it does not wish to be seen it may dive from a height of say 1,000 feet into the wind, checking its fall some distance behind and slightly below its prey. It then corrects the line of attack and thrusts upwards with closed wings, lashing out with its talons at the underside of the victim's body. The trajectory is so angled as to be invisible to the bird under attack.

Another stratagem frequently employed with a following wind is for the falcon to dive down at an angle but this time to check the descent some way in front of the victim. When it is no more than 30–40 feet from the oncoming bird's flight path the raptor does a smart about-turn in mid-air and swoops down to collide almost head-on with its prey. In such cases the falcon's aim must be absolutely true for if it fails to deal a telling blow at head or wing it may sustain serious injury from the force of the impact, given the combined velocities of attacker and attacked.

Whatever the mode of attack the peregrine falcon has to ensure that in dealing the initial blow it does not run the risk of breaking its own leg. Because of the speed at which it launches the attack the important point is not so much the force of the blow as its accuracy. Thus the falcon's talons may barely graze the other bird's body but this will usually be sufficient to rip the flesh open and possibly reach a vital organ.

Hunters of field and plain

Given these various methods of attack it is understandable that the peregrine falcon should prefer to pick its victims from those species that normally fly quite fast in a more or less straight line – birds such as pigeons, teal, sand-grouse, petrels, guillemots and the like. Slower-flying species with somewhat more acrobatic talents, such as the crow family, gulls and lapwings are less frequently challenged. The reasons are several. The direct, high-speed mode of attack favoured by the falcon is most effective when directed against a bird which because of its own speed has little chance to practise evasive manoeuvres. Another advantage of attacking species that engage in rapid wing-beating flight is that their muscles are well developed and their bodies thus rich in protein.

Peregrine falcons require plenty of open ground for hunting, whether it be steppe, prairie, cornfield or other type of arable land. It is essential that any trees should be well separated and not form compact wooded clumps. Many of them catch their prey over the sea and carry it back to shore, where they eat it. Yet despite their inability to function effectively against prey sheltering in trees and bushes or hiding in grass and corn, the raptors are not entirely absent from forested mountain areas, though they cannot be said to be numerous in these parts. In Spain, for example, they are sometimes sighted in hilly districts, as in the Basque provinces, Santander, Asturias and Galicia. Some falcons even frequent olive groves where fieldfares are their favourite victims; but those that winter in the meadows of Estremadura and among the pine woods of the central provinces prey principally on rock doves.

Broadly speaking, pigeons and doves of various species form the basis of the peregrine falcon's diet the world over, except for those living in the Volcano Islands and the Queen Charlotte archipelago, which feed mainly on sea birds. The falcon population of Spain is especially dependent on domesticated and wild pigeons, but as the following survey by Felix Rodriguez de la Fuente shows, the choice of food varies from season to season.

Birds for all seasons

It has been estimated that between two-thirds and three-quarters of the food consumed by a peregrine falcon normally consists of different species and varieties of pigeons. Nevertheless, the following survey of the feeding habits of the bird in Spain reveals interesting variations in diet and behaviour directly related to the changing phases of its biological cycle; and these modifications naturally have important repercussions on the life patterns of other bird species.

1. Courtship display. Peregrine falcons are sexually active from the beginning of February until the end of May (the period varies according to region), during which time the significant modifications in their behaviour effect many species that are normally undisturbed by the raptors.

Analysis of remains of prey found on territories of twenty pairs of sedentary peregrine falcons over five-year period

Frequent prey – all seasons

Domesticated rock dove (*Columba livia*)
Wild rock dove (*Columba livia*)
Wood pigeon (*Columba palumbus*)

Frequent prey, varying with season and region

Turtle dove (*Streptopelia turtur*)
Spotless starling (*Sturnus unicolor*)
Starling (*Sturnus vulgaris*)
Skylark (*Alauda arvensis*)
Crested lark (*Galerida cristata*)
Calandra lark (*Melanocorypha calandra*)
Jackdaw (*Corvus monedula*)
Lapwing (*Vanellus vanellus*)
Redwing (*Turdus musicus*)
Mistle thrush (*Turdus viscivorus*)
Great spotted cuckoo (*Clamator glandarius*)

Infrequent prey

Black-bellied sand-grouse (*Pterocles orientalis*)
Teal (*Anas crecca*)
Stone curlew (*Burhinus oedicnemus*)
Snipe (*Gallinago gallinago*)
Stock dove (*Columba oenas*)
Red-legged partridge (*Alectoris rufa*)
Little bustard (*Otis tetrax*)
Quail (*Coturnix coturnix*)
Linnet (*Carduelis cannabina*)
Swift (*Apus apus*)
Green woodpecker (*Picus viridis*)

Very infrequent prey

Rock sparrow (*Petronia petronia*)
Mallard (*Anas platyrhynchos*)
Great spotted woodpecker (*Dendrocopos major*)
Golden oriole (*Oriolus oriolus*)
Blackbird (*Turdus merula*)
Great reed warbler (*Acrocephalus arundinaceus*)
Great grey shrike (*Lanius excubitor*)
Hoopoe (*Upupa epops*)
Cuckoo (*Cuculus canorus*)
Little owl (*Athene noctua*)
Red-necked nightjar (*Caprimulgus ruficollis*)
Magpie (*Pica pica*)
Greenfinch (*Carduelis carduelis*)
Kestrel (*Falco tinnunculus*)
Moorhen (*Gallinula chloropus*)
Black redstart (*Phoenicurus ochruros*)
Black wheatear (*Oenanthe leucura*)
Raven (*Corvus corax*)

a) At this time of year the males stake out territories and join battle with rivals and intruders of the same species and sex, such combats sometimes being fought to the death.

b) They expel from the rock ledges chosen as nesting sites hordes of jackdaws (*Corvus monedula*) that settle in the area, forcing them to seek refuge on rocks and in woods at lower altitudes.

c) Jackdaws are often killed in the course of the male falcon's courtship display, for a feature of this performance is the capture of birds not normally hunted, these being offered to the female.

d) Towards the end of February and during March–breeding season for many species of birds–most of the dead victims examined for the purposes of the survey proved to be males. This discovery was independently confirmed by direct study of the peregrine falcons' hunting methods. It was observed that they frequently captured pigeons and calandra larks while the latter were engaged in nuptial flight displays. Trained falcons also instinctively attacked species that in the ordinary way they would not touch. These findings would seem to lead to the conclusion that the exhibitionistic behaviour and ornamental plumage of the reproductive males at breeding time (contrasting so markedly with the reticent habits and relatively drab colours of the females) readily attract the attention of their principal predators, whereas their partners, already fertilised, are spared.

e) Examination of the feet from the corpses of five red-legged partridges showed that four were equipped with spurs, indicating that they belonged to males. It would appear that the latter, exhausted by their ritual combats, had managed to offer little resistance to the falcons, presumably having no energy left even to flatten themselves against the ground for self protection.

f) An extremely wide range of prey is evident during the peregrine falcons' breeding season, partly because it coincides with the return of migrating species and partly because of the males' tendency to offer unaccustomed prey to their partners.

2. Egg-laying and incubation. In central Spain the female peregrine falcon lays 3–4 eggs at the beginning of March and incubates them for about 28 days. During this time she never leaves the nesting site and is fed by her mate. There is now a significant change in food habits.

a) Birds of small or medium size–starlings, larks, lapwings and the like–are frequently captured, these being well suited to the hunting capacities of the male falcon which is lighter in weight and more agile than the female.

b) There are appreciably fewer species in evidence now than at the time of the nuptial display.

c) Indications are that the incubating bird eats very little since she does not move about and requires fewer calories than at other times.

3. Rearing of fledglings. The fledglings do not move from the eyrie for about five weeks. The male continues hunting and the female tears up the prey into suitably sized portions. She also defends the brood from enemies.

In order to get as accurate an idea as possible of the food habits of the family during this period, the members of the study group chose a vantage point on a hill overlooking a wide valley with trees and cultivated fields. Their hide was only 10 feet or so from a nest occupied by four fledglings. The observers were able to study the birds from dawn to dusk and to record all the family activities. The experiment was carried out three times, each at a different stage of the fledglings' development.

During the entire period of rearing the young, food consisted in the main of domesticated or wild pigeons. The young falcons appeared to consume a little in excess of one pigeon every day but this quantity of food diminished the nearer they came to reaching full growth. The combined food intake of the adults was also the equivalent of one pigeon per day. The male, smaller than his mate, evidently had a more modest appetite. All the birds he captured were comparatively small. The mother would eat part of the prey herself and after that she would offer what was left to her offspring.

4. Hunting initiation. As soon as the young are capable of leaving the eyrie–generally during May–up to the time they wander off to lead an independent life (July–August) the parents introduce them to their hunting grounds and teach them how to fend for themselves. This apprenticeship period sees yet another modification of food habits, with direct consequences for many birds

If the prey is not killed outright in mid-air it is immediately seized and dragged to the ground, then dispatched with a blow of the bill to the base of the neck.

Facing page: The kestrel, by rapid, rhythmical beating of its long wings, can hover almost motionless in the air in order to survey the terrain below. Like other falcons it does not build a nest but frequently appropriates one originally belonging to another species.

KESTREL
(Falco tinnunculus)

Class: Aves
Order: Falconiformes
Family: Falconidae
Length: 12½–14½ inches (31–37 cm)
Wing-length: 9–10½ inches (23–27 cm)
Wingspan: 27–32 inches (68–82 cm)
Weight: 6–8 oz (180–230 g)
Diet: small mammals, insects, reptiles, small birds
Number of eggs: 3–7, usually 5
Incubation: 27–31 days

Back reddish-brown with dark brown spots, throat white, chest creamy with brown spots. Male's head, rump and tail are grey. Female's tail is reddish-brown with narrow black bars. Both sexes have a large black band, fringed with white, at the tip of the tail. Iris dark brown, cere and feet yellow.

LESSER KESTREL
(Falco naumanni)

Class: Aves
Order: Falconiformes
Family: Falconidae
Length: 11–13 inches (27–33 cm)
Wing-length: 9–10 inches (22–24·5 cm)
Wingspan: 24–26 inches (61–66 cm)
Weight: 4–5 oz (120–145 g)
Diet: insects, small lizards and other vertebrates
Number of eggs: 4–5
Incubation: about 28 days

Smaller than the kestrel, the moustachial stripe is faint, the claws are white and there are no spots on the back although the female's back has narrow black bars. Cere and feet yellow.

of small and medium size.

a) As spring turns to summer the immature falcons are taught how to capture baby birds just learning to fly.

b) In the course of these semi-playful hunting expeditions the young falcons harry jackdaws, sometimes killing them and often causing such disturbance to the incubating females that many of their eggs fail to hatch.

c) A number of remains of red-legged partridges were found–apparently injured or undeveloped individuals–whose feathers had been removed by the immature raptors.

d) During this period a greater number of victims are to be found than at other seasons but not a large variety of different species.

5. *Parting of young from adults.* In autumn and winter the standard rock dove diet of the peregrine falcon is supplemented by starlings, larks and other species that arrive to spend the cold season in Spain. In fact the sedentary falcons cut down on their consumption of these pigeons and leave them for the most part to the visiting migrants from the north, stepping up their own attacks on little bustards, ducks and small wading birds. By this time the partridges have reached adult size and are too large to be caught. Few remains of this species are found. The jackdaws leave their rock ledges and crevices to assemble in large flocks but no longer attract the predators. Thus during the winter the range of prey is much more restricted than it tends to be at other seasons.

6. *Conclusions.* The day-to-day observations of the feeding requirements of trained falcons, coupled with surveys of wild falcons over a period of years and examination of their food remains, have enabled Spanish ornithologists to draw the following general conclusions about the influence of the peregrine falcon on local birdlife.

a) An adult falcon consumes daily a quantity of protein equivalent to a pigeon but does not trouble to catch additional prey once its appetite is satisfied.

b) About 75 per cent of all food eaten by the Spanish subspecies consists of different races of pigeons, both tame and wild.

c) Among the smaller birds habitually eaten are starlings, turtle doves and, to a lesser extent, skylarks, short-toed larks and calandra larks.

d) Partridges are seldom attacked by peregrine falcons which are not expert in hunting on or near the ground. Only in the spring are a few males and fledglings killed. Incidentally the raptor has a beneficial effect on the reproductive cycle of the latter species for the males that do succumb are generally those exhausted by intraspecific fighting whilst defending territory and trying to win a partner.

e) The immature falcon hunts jackdaws when first learning to capture prey and the adult bird attacks them in the course of the latter's courtship displays. When breeding, the falcon expels jackdaw colonies from rocky regions and drives them down into

woods where they are easily shot by game wardens concerned to control their numbers.

f) The comparatively high density of the peregrine falcon population in Spain is due to favourable topographical and climatic conditions and to the abundance of pigeons. The latter are prevented from becoming too numerous and damaging crops.

g) From the viewpoint both of hunters and farmers, the peregrine falcon must be regarded as a useful, not harmful, bird.

Other falcons of the plains

Before pesticides were introduced and when hunting was not as popular a sport as it is today, the Castilian plains were a paradise for many other members of the falcon family, including the kestrel (*Falco tinnunculus*) and the lesser kestrel (*Falco naumanni*), small raptors feeding on insects, rodents, reptiles and birds. They can still be seen here and there, perched on a fence or telegraph pole, and are notable in flight for the way in which they hover to survey the underlying terrain. The lesser kestrel is of course the smaller, identifiable by its white claws. It nests in colonies–sometimes on buildings–and is migratory, spending the spring and summer in Spain and flying off in autumn to winter in Africa. The kestrel is a stronger bird, with black claws. The population is mainly sedentary and in central provinces this raptor performs a service to farmers by destroying many rodents. When the latter seek safety underground the kestrel turns its attention to birds.

The merlin (*Falco columbarius*) is another small falcon that lives in Spain from the end of September until March, nesting in northern Europe and the British Isles. This winter visitor to the Mediterranean–which looks like a miniature version of the Arctic gyrfalcon–hunts mainly birds, especially larks.

Other birds of prey ideally adapted to these southern cereal-growing flatlands are the various harriers, notably the hen harrier (*Circus cyaneus*) and Montagu's harrier (*Circus pygargus*). At one time these species confined themselves to open uncultivated country but nowadays they conceal their nests in fields of corn and barley, feeding plentifully on insects, rodents, reptiles and amphibians. They also concentrate their attacks on the fledglings of ground species. The males are attractive with grey plumage and vividly contrasted black wing-tips, their bodies clearly silhouetted against the gold of the summer corn or the brown winter stubble. The slender wings and long tail give the males a graceful outline in flight. In some parts of Castile, as in the Tierra de Campos, harriers are probably the commonest local birds of prey and very valuable to farmers.

The peregrine falcons and harriers are joined in spring and summer by another small raptor, the hobby (*Falco subbuteo*), which nests in trees and hunts insects and small birds.

Ranging from tiny beetles (favourite food of the lesser kestrel) to the little bustard (speciality of the peregrine falcon), there is no animal in these parts, whether it be insect, frog, snake, lizard, mouse or bird, which is not the potential victim of one or more of these predatory falcons.

Facing page: The hobby is a small falcon which nests on the fringes of woods and forests, hunting insects and small birds. By the end of summer the fledglings are sufficiently developed to fend for themselves and make ready for the autumn migration.

HOBBY
(*Falco subbuteo*)

Class: Aves
Order: Falconiformes
Family: Falconidae
Length: 11–14$\frac{1}{2}$ inches (28–36 cm)
Wing-length: male 9$\frac{1}{2}$–11 inches (24–28 cm)
female 10$\frac{1}{2}$–12 inches (26·5–30 cm)
Wingspan: 28$\frac{1}{2}$–33 inches (72–84 cm)
Weight: 5–12 oz (150–340 g)
Diet: insects, birds, rarely small mammals and reptiles
Number of eggs: 2–4, usually 3
Incubation: 28 days

Top of head and moustachial stripe very dark brown, almost black. Throat, sides of neck white. Back slate-grey, chest light brown with black streaks. Belly and legs reddish-brown. Iris dark brown, cere and feet yellow.

MERLIN
(*Falco columbarius*)

Class: Aves
Order: Falconiformes
Family: Falconidae
Length: 11–13$\frac{1}{2}$ inches (28–34 cm)
Wing-length: 7$\frac{3}{4}$–9$\frac{1}{2}$ inches (19·5–23·5 cm)
Wingspan: 22–27 inches (56–69 cm)
Weight: 5–9 oz (150–260 g)
Diet: almost exclusively birds but occasionally small mammals and insects
Number of eggs: 3–6, usually 4
Incubation: 28–31 days

Male smaller than female. Back blue-grey, throat and sides of head whitish, chest light red with brown spots. Tail grey, with a black-barred, white-fringed tip. Female's back dark brown, tail feathers barred with white. Iris dark brown. Cere and feet yellow.

Caracara
(*Caracara cheriway*)

Peregrine falcon
(*Falco peregrinus*)

Barred forest falcon
(*Micrastur ruficollis*)

African pygmy falcon
(*Polihierax semitorquatus*)

The Falconidae comprise three subfamilies. The Herpetotherinae include the barred forest falcon, and the Polyborinae the South American caracara. The Falconinae have a world-wide range and among its well-contrasted representatives are the African pygmy falcon and the peregrine falcon.

FAMILY: Falconidae

In the order Falconiformes the family Falconidae, which contains the falcons and their relatives, is made up of almost 60 species, 40 of which are falcons in the true sense, grouped in the genus *Falco*.

Many ornithologists regard the falcons as being quite different from other diurnal birds of prey, so much so that they would seem to have evolved independently, justifying separate classification as the order Falconiformes. Other diurnal birds of prey, in the view of these authors, should be classified as Accipitriformes.

Certainly the Falconidae and Accipitridae are significantly different in many aspects of anatomy, physiology and behaviour. Falconers recognise these differences by employing separate methods of training and by using the birds for the types of hunting for which the two families are best suited.

The head of a typical falcon is rounded, the line of the brows being only faintly visible and the eyes dark. This gives it a rather less fierce expression than that of other raptors. The second wing feather is generally somewhat longer than the rest, whereas in the case of the Accipitridae it is the fourth wing feather which stands out. The upper mandible is characterised by a 'tooth', not present in other diurnal birds of prey, although eagles possess a curved edge to the upper mandible which makes it especially sharp.

There are obvious contrasts too in reproductive behaviour. Falcons do not construct a nest but rear their young either in the abandoned nest of another species or on bare ground. Some species choose nesting sites on high, unprotected rock ledges.

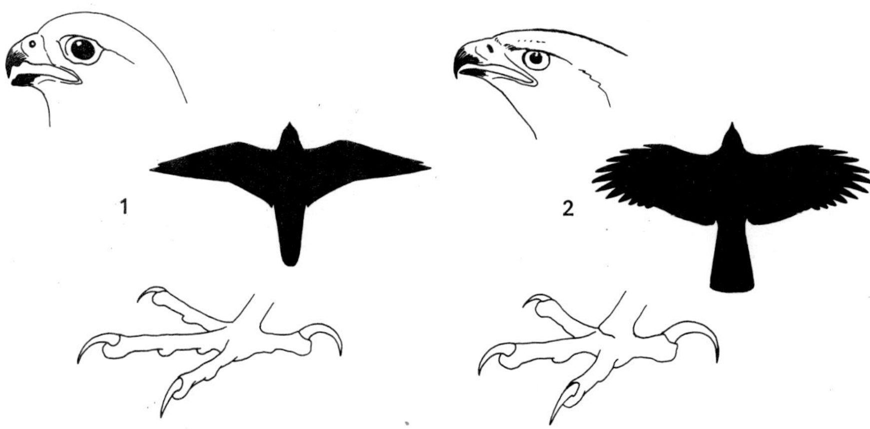

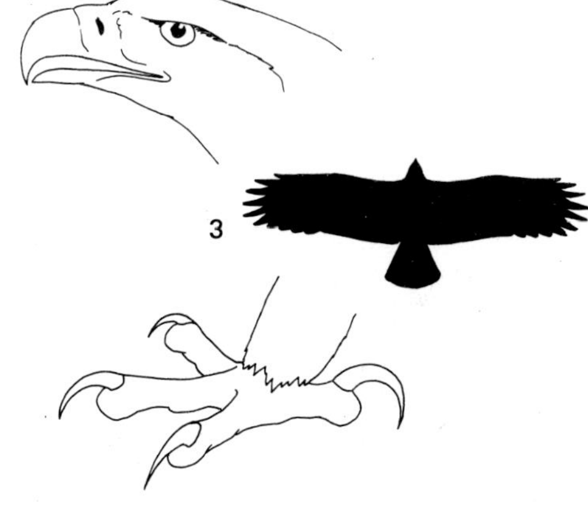

Diurnal birds of prey, employing a variety of hunting methods, are anatomically distinguishable by their silhouette in flight, the shape of head and bill and the structure of feet and claws. These falcons (1) have a streamlined body, pointed wings, long claws, a rounded head and a 'toothed' upper mandible. They feed mainly on other birds captured on the wing. Goshawks (2) have broad, rounded wings, a long head and medium-size claws–suitable for gliding at low level among trees and capturing all types of prey. Eagles (3) have very long wings, a long and large bill and short but powerful claws, making them remarkably good fliers and formidable predators with a wide range of prey.

When they venture close to water the members of the family Accipitridae wet their wings by simply beating them against the surface. The Falconidae, on the other hand, actually splash their wings in the water.

The standard classification of these diurnal birds of prey into two families, Accipitridae and Falconidae, grouped under the single order Falconiformes, can perhaps be justified by pointing to the apparent resemblance of some species that could be described as primitive falcons to harriers and other Accipitridae. Thus the representatives of the subfamily Herpetotherinae (genera *Micrastur* and *Herpetotheres*), known as the forest falcons and laughing falcons of South America, look like a cross between harriers and vultures, although they possess many typical falcon features and do not build their own nest. These are clearly ancient forms which have adapted to hunting in the thickets of tropical forests and have evolved differently from other members of the family.

The subfamily Polyborinae are commonly known as caracaras, some of which are insectivores, others omnivores. They too are South American birds, but construct their own eyrie, and some scientists refuse to regard them as true Falconidae. They are massive birds with long feet, somewhat resembling buzzards in their flight silhouette. But some of them have a dark brown iris and rounded nostrils–characteristics of falcons.

The subfamily Falconinae comprises falcons proper, with a world-wide distribution from the northern Tundra to the burning desert. Their prey ranges from tiny insects to birds as large as bustards and geese. Some of them are specialised hunters of the skies, such as the peregrine falcon with its 200 mile-per-hour dive. The projections on the nostrils of these predators may help them to breathe as they plummet through the air.

All falcons are extremely powerful fliers. They have narrow, pointed wings and a broad chest. All the senses, particularly vision, are very well developed. The reason that a saker, for example, cannot spot a bustard over a distance of more than about nine miles is not due to poor eyesight but to the curvature of the earth's surface!

The Falconidae have many characteristics in common but are nevertheless extremely diversified. They do not therefore form a homogeneous group as does, for example, the genus *Accipiter* (goshawks and sparrowhawks) and for this reason the forty or so representatives of the genus *Falco* are often divided into various subgenera.

Many fossil remains of Falconidae have been discovered dating back to the Tertiary period. Traces of the genus *Falco* have been found in Early Miocene layers in Nebraska and of the genus *Badiostes* in Argentina. Falcons with long spindly legs (caracaras) dating from the Early Pleistocene have been excavated in Texas and Florida; and other fossils from California, New Mexico, Florida, Mexico, the Bahamas, Puerto Rico, Brazil and Ecuador go back to the Late Pleistocene. No fossil remains have been found of any forest falcons, but bones of pygmy falcons have been excavated in Brazil which also date from the Late Pleistocene.

CHAPTER 7

The changing face of the deciduous forest

Life in the modern technological age is becoming ever more restrictive, programmed and stereotyped. Traffic-congested highways link sprawling cities which more and more resemble depersonalised jungles of concrete, metal, asphalt and glass. Inured to fog, smoke, fumes and noise, the town dweller longs for the tranquillity, clean air and calm beauty of the countryside, surroundings in which he can find refreshment for body and mind. Woods, fields, moors, mountains, lakes and rivers offer welcome respite from the pressures and cares of everyday life.

At a deeper level nature evokes a mystical sense of wonder and stimulates the creative sensibility. The leaf is a miracle, the tree an image of grace and strength, the forest a symbol of permanence. Each season casts its individual spell of beauty–spring with its burgeoning shoots, summer with its shady foliage, autumn with its crisp falling leaves, winter with its stark contrasts of black and white. The changing face of the forest presents a panorama of natural form and colour, unsurpassed in splendour and majesty.

Deciduous woods and forests are found both in the Old and New Worlds, growing wherever there is adequate rainfall and a suitably temperate climate. The latter requisite, however, shows what a false impression can be derived from cold statistics. Judging by average temperature and rainfall, central Europe, Manchuria and the eastern seaboard of the United States are all classified as temperate zones, yet in all these regions winter may be as cold as in the frozen Arctic while in summer temperatures may be recorded as high as anything experienced in the tropics. Furthermore, there may be equally dramatic seasonal variations in the rainfall pattern.

Given such striking contrasts in climate, local flora and fauna

Facing page : The deciduous forest in autumn is a riot of glowing colours–offering shelter and food for a multitude of animals and providing an atmosphere of calm beauty for nature lovers.

Silviculture was developed in Europe in the 19th century with the object of preserving and restoring the woodland environment. Many deforested areas have since been replanted but unfortunately the broad-leaved deciduous species that were characteristic of the primeval forest have been sacrificed in favour of faster-growing and more profitable conifers.

are obliged to adapt themselves to rapidly changing conditions. It is probably for this reason that the deciduous forest, responding in such a diverse and spectacular fashion to outside stimuli, gives the impression of being a habitat that is always teeming with life, whatever the season. Desert, steppe and coniferous forest, by comparison, seem static, almost petrified, apart from a brief carpet of greenery or a scattering of bright flowers in the spring.

The characteristic feature of a deciduous tree is the autumnal shedding of its broad leaves–a functional response to the impending rigours of winter. Leaves are the tree's transpiratory organs from which large quantities of water are evaporated. All plants need water to survive and the most efficient method of reducing liquid loss (for even if there is water in winter it is often in the form of ice and cannot be utilised) is to dispense with these special organs. The bare branches of the tree store moisture in the same way as do the stiff spines of a desert cactus. With the arrival of the spring rains, followed by humid summer heat, the tree again needs a broad surface of evaporation and the bare branches are once more festooned with foliage.

The changing face of the deciduous forest is a natural phenomenon in response to climatic modifications, comparable to the migratory journeys of flocks of birds, the seasonal moulting of feathers or shedding of skin, the winter lethargy of insects, the hibernation of cold-blooded (and some warm-blooded) vertebrates, and a host of other changes in the appearance, physiology and behaviour of living creatures.

Man and the forest menace

At one time deciduous forests covered a vast part of the Holarctic region. In Europe they spread from the British Isles eastward to Poland and Russia, and from Sweden southward to Aquitaine. In the eastern half of the United States they stretched from the Great Lakes down to the Gulf of Mexico. In eastern Asia too they covered an enormous tract of land which included Manchuria, Korea and Japan. Furthermore, between the steppes of central Asia and the coniferous forests of the nordic taiga there existed a broad, vaguely defined belt of tree-steppe, made up for the most part of birches and willows.

Nowadays there are only isolated remnants of this vast area of woodland. The reason for its disappearance are several, but whether directly or indirectly caused, man has been the culprit. Whatever his motivation he has come near to exhausting some of the most valuable natural resources of his planet.

In ancient times–with the exception of the Mediterranean countries–the forests of Europe consisted chiefly of oaks and beeches, trees that normally provide a home for an extraordinarily rich fauna. Julius Caesar in his famous *Commentaries* described the impenetrable forest of central Europe, a savage and mysterious place inhabited by barbarian tribes. Caesar estimated that it would take nine days to travel across the tree-covered mountain range in Germany known as Hercynia Silva. To cross it from west to east from the point where it started on the borders of the land of the Helvetii (western Switzerland) to its easternmost limits far down the Danube was reckoned to be a sixty-day journey.

This forest originally harboured many animals–bison, aurochs, deer, wild boar and bear–and was evidently associated in the civilised Roman mind with the primitive, hostile tribes that were continually hammering away at the Empire's frontiers. This view perhaps provides a clue to later events. For all the mundane, practical reasons put forward, it may have been this traditional conception of the forest as a dark, menacing, evil place that subconsciously inspired man to pit his will against it and do his utmost to destroy it.

Within the last few centuries the conflict between old and new has erupted wherever colonists have clashed with traditional, indigenous patterns of life. In tropical Africa, North and South America and Australia, white settlers have hacked down trees and cleared land to make way for their farms and plantations, callously uprooting tribes that have for centuries depended on the forest for their existence.

The feelings of the newcomers towards these primitive peoples have been compounded of fear, hostility and contempt. The forest has represented a natural barrier to progress and profit. Similar sentiments may well have motivated those who

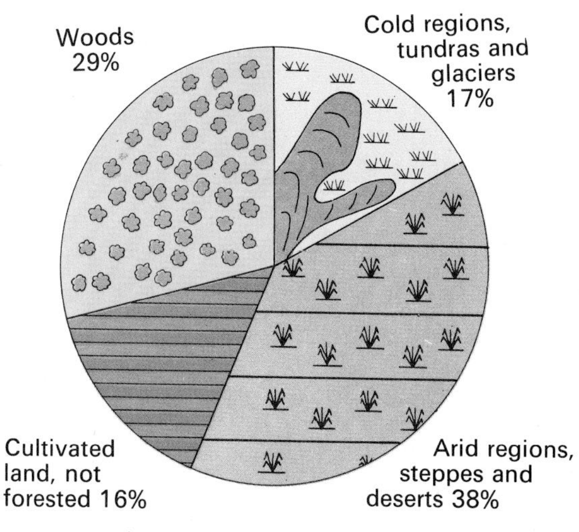

This diagram by the French ecologist Viers shows the comparative importance of forest and woodland in relation to the earth's total land area.

Four hundred years ago the North American deciduous forest was one of the largest and most varied in the world. Today, according to Shelford, not more than one per cent of this forest remains.

For centuries the forest was regarded as a wild, mysterious and menacing place and civilised man had no compunction in pushing back its frontiers. It is only in recent times that he has become aware of the danger involved in thus upsetting nature's equilibrium and has begun to try to make amends for the previous devastation.

initiated the processes of deforestation in Europe and Asia. Granted the growing demand for timber as building material and fuel, practical considerations alone could not possibly have accounted for the systematic devastation of entire forest regions.

What remained of the primeval forest was eventually tamed. In the 19th century Europe saw the development of the science of forestry or silviculture, designed to protect and restore the woodland environment. Artificial forests were planted, using selected tree species which could be expected to give a good timber yield. Beech and oak – principal constituents of the ancient forest – were often sacrificed, for economic reasons, in favour of quicker-growing, more profitable conifers, though happily fragments of original forest and woodland were preserved here and there in central Europe. Many animal species had by this time vanished but others were saved from extinction; and many insectivorous birds continued to find food and shelter among the trees.

The same fate was in store for the forests of North America but here the destructive process was far more rapid. At the beginning of the 17th century only a few areas of woodland had been touched, notably in Virginia with its large Indian population. Damage to forests was mainly due to fires, accidentally caused or deliberately lit by the Indian tribes in a bid to hem in the huge buffalo herds.

Today it has been estimated that barely one per cent remains of the New World's primeval forest. Much of the land that was cleared was successfully given over to crops and livestock but in many areas poor farming techniques and adverse weather conditions ruined the soil, turning fields into dust-bowls. In many regions the damage was irreversible; elsewhere the rampant progress of erosion was checked by planting new trees.

The wolf, the puma and the wapiti, driven from their original habitats, vanished from many districts. On the other hand, the changed conditions suited other species admirably. The black bear and the wild turkey found the plant composition of their new environment so much to their liking that their numbers soon rose dramatically.

Somewhat belatedly man has realised that the forest is more than just a refuge for certain animal species and that it is in fact part of nature's grand plan. It is no exaggeration to say that man's survival is at stake. The French ecologist Viers has pointed out that the forest is no longer a threat to civilisation but the very reverse – a potential life-saver for millions of people. All the more important, therefore, that these islands of greenery should be preserved as among the most precious legacies bequeathed to us by our ancestors.

If we continue to destroy our deciduous forests, which are almost as old as the earth sustaining them, we are not only bound to decrease the productivity of the soil but by interfering with the natural balance of the environment run the real risk of endangering our future. Commonsense demands that our remaining oak and beech forests should be protected, not put to the axe and then replaced by conifers. The latter should be planted on existing tracts of open land, together with such

ECOLOGICAL PYRAMID OF VERTEBRATES OF THE DECIDUOUS FOREST
SUPER-PREDATORS
Golden eagle
Wolf
Eagle owl
PREDATORS
Sparrowhawk
Tawny owl
Goshawk
Buzzard
Barn owl
European lynx
Marten
Brown bear
Badger
Polecat
Weasel
HERBIVORES
Crossbill
Blackbird
Fat dormouse
Thrush
Red deer
Squirrel
Vole
Capercaillie
Roe deer
Wild boar
Bullfinch
Hare
Partridge
Beech
Lime
Grey poplar
Oak

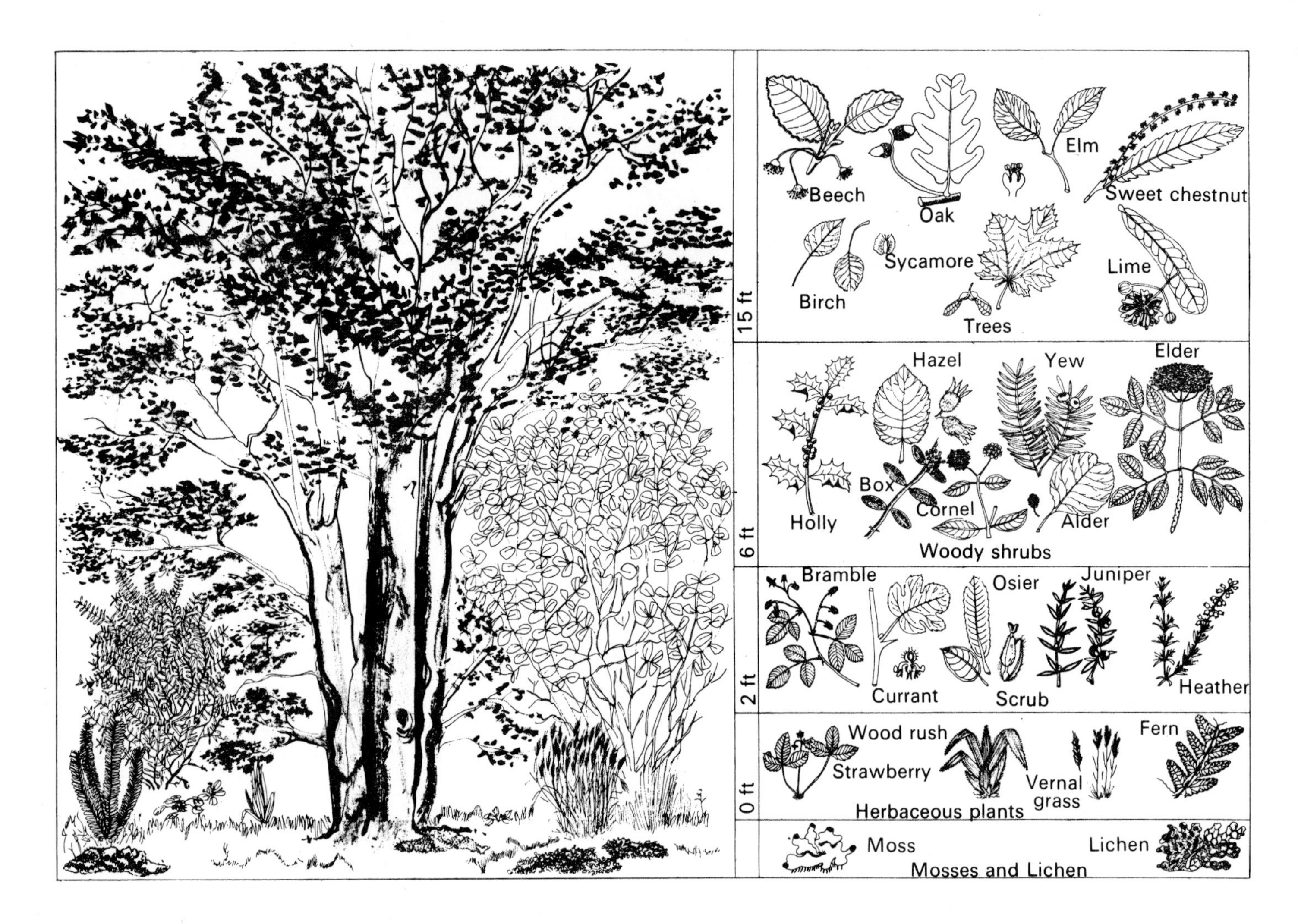

Most botanists subdivide the vegetation of the deciduous forest into three levels–trees, woody shrubs, and herbaceous plants. Others add two more–scrub and mosses and lichens. The five-level structure is shown in this diagram, with appropriate plant species.

species as eucalyptus, which can be utilised by industry.

It is difficult to trace the precise boundaries of the deciduous forest–except where it is fringed by ocean–because it appears in so many different guises and gives way imperceptibly to other natural biomes. In Europe, where it is notable for a comparative poverty of plant species, it is roughly bounded on the north by the taiga, on the south by the Mediterranean maquis and on the east by the herbaceous steppe. In Asia it is likewise fringed to the north and the north-east by the taiga, to the west by the lofty counterforts of the high plains of central China and to the south by the alluvial plains of the Yellow River and the subtropical forest which once covered the whole vast region of southern China. In the United States, where ecologists describe the deciduous forest as the 'oak-deer-maple biome', there are three clearly recognisable types of woodland. In the north and on high ground in general maple, oak and chestnut predominate; in the south and on low-lying terrain the principal species are oak, hickory and magnolia; and along streams and rivers there is a mixture of all these species.

The North American forest is bounded to the west by the Great Plains which stretch as far as the Rockies. In the north, as in Europe and Asia, it gives way gradually to coniferous forest. In the south it runs down to the shores of the Gulf of Mexico on one side and on the other merges with the subtropical forest which covers the tip of Florida.

Beech
Oak
Sycamore
Lime
Ash
Birch
Elm
Poplar
Sweet chestnut

Trees and shrubs of the deciduous forest

The most characteristic trees of the deciduous forest are beech and oak. The former, very typical of wet regions and chalk soil, forms an enormous curtain of green in the summer through which very little sunlight filters. Consequently hardly any plants can grow under this canopy of leaves and the ground in a beech wood is bare of greenery for most of the year. The latter, more adaptable and with a number of species, is widely distributed through Europe, Asia and America. In the New World about 85 per cent of deciduous trees are oaks. They burst into leaf comparatively late in spring and drop their leaves after other trees in autumn. Apart from these two magnificent types, the trees most commonly found are maple, lime, chestnut, ash, hazel, birch and elm. The magnolia grows in the southern part of the United States along the so-called maritime fringe, where climatic conditions are almost subtropical.

Trees are not of course the exclusive constituents of the deciduous forest which should properly be regarded as an entity consisting of a number of different elements at varying levels. The uppermost level is of course made up of tall trees, ranging in height from 20–100 feet or more; then there is an intermediate level of vegetational growth comprising trees and shrubs up to about 20 feet high; finally, at the lowest level, there are bushes, grasses, flowers, dead leaves and humus (decomposed leaves and other vegetable and animal matter).

Among characteristic species at the intermediate level are cornel (which turns red in winter), apple, pear, mountain ash, briar and bramble, together with honeysuckle and other climbing plants that wind their way around tree trunks. The holly may sometimes grow to a height of 30 feet but is normally less.

The dead leaves and organic humus at ground level is interspersed with grass and low bushes. Some species, such as the violet, flower before the trees are covered with foliage, while there is still a chance for the sun's rays to break through. Other sturdy species of the undergrowth include ferns, heather, cytisus, gorse and bilberry.

Some authors subdivide the forest layers into five levels. The topmost level comprises trees that stand above 15 feet; then comes a layer of smaller trees and shrubs, and a third consisting of thickets of about the same height as the previous group but made up of many different species. The herbaceous level is of course the grass cover and the fifth and final layer consists of moss, lichen and fungi.

The vegetational composition of a forest may fluctuate from season to season. In a beech wood, for example, there will be only one layer of growth in summer or winter–the arborescent one–with multiple levels in spring and autumn when the ground is dotted here and there with patches of greenery, small flowering plants, ferns and fungi.

It is easy to theorise about the nature and composition of a deciduous forest but the reality is often quite different, as will be appreciated by anyone who has walked through or flown

over an area of woodland. In fact most of these forests are a blend of deciduous and coniferous species, with pines, firs, Norway spruce and cedar well represented. In Poland some 95 per cent of trees in what are termed deciduous forests are conifers. The figure for Germany is 60 per cent, for France 50 per cent. The reason for this phenomenon is that the deliberate planting of rapidly growing coniferous species leads to changes in the nature of the soil so that areas denuded of vegetational cover are soon taken over by these fast developing trees. After axe and fire have destroyed the broad-leaved species, the humus formed by the needles of conifers that replace them acidify the subsoil, preventing the growth of beech, oak and other trees that require alkaline conditions. Once begun, the process is difficult to reverse as the soil becomes less and less conducive to the development of the traditional species. Thus the felling of venerable oaks and their replacement by pines or eucalyptus is in a sense an assault on nature.

In many parts of the world, therefore, the deciduous forest is, strictly speaking, a mixed forest of broad-leaved trees and conifers. It is nowhere more varied than in eastern China's Chekiang province where 180 species have been counted, 67 of them very tall. The flora of the North American forest is also rich, with almost 40 different species of oak. By comparison the forests of Europe are threadbare. The natural barriers formed by the

The European bison or wisent was once the most common of forest animals. The last wild bison disappeared at the beginning of the 20th century but attempts have since been made to reintroduce the species, using animals reared in zoos. For several years a small herd of wild bison has been breeding in the Bialowiecz forest between Poland and the Soviet Union.

Facing page : Some trees in the deciduous forest, such as the poplar (*above*) and the willow, grow on the shores of rivers and streams. Others are better able to stand drought and flourish away from water. Most deciduous forests contain conifers (*below*) as well as characteristic broad-leaved trees.

Sahara and the Mediterranean have blocked the northward spread of tropical species. Viers has pointed out that not counting waterside species such as willow and poplar and a few rare, isolated woodland species, there are not more than twenty common trees in France, of which eight are conifers.

The living forest

There is probably no other natural habitat which offers a more secure refuge to its animal inhabitants than the deciduous forest. Below ground, among the dead leaves that strew the surface, in the thickets, between the barrel of the tree and the bark, in roots and hollows and on high branches, innumerable living creatures shelter, feed and breed. Fosco Mariani has described the forest as a living community rich with fruits, flowers and animals, an environment which pulsates with mysterious whispers, bubblings, murmurs and rustlings.

The leaf litter swarms with bacteria, fungi, tiny arthropods, worms and protozoans, which transform the dead vegetation into humus and annually refertilise the soil. Buried in the decaying leaves are mites, small scarabs, woodlice, springtails, rove-beetles, rotifers and centipedes; and above the surface are snails, molluscs and ants. These animals serve as food for many vertebrates including toads, salamanders, lizards, blindworms and mammals such as shrews and hedgehogs. The mole has little to fear from competition for food and enjoys relative security from predators, in contrast to small animals foraging for insects and worms at the surface.

Among the carnivores of the forests, prowling at ground level, are various snakes, badgers, weasels and polecats (beech and pine martens being found more frequently higher up in the trees). Larger predators include foxes, wolverines, wildcats (they too hunt overhead), and even bears. Additionally, in the North American forests there is the Virginian opossum, only marsupial of the Holarctic region.

Anything at surface level is potential food for the wild boar which, with the aid of its prominent snout, scoops insects from the subsoil as well as feeding on vegetation. The red deer, roe deer, fallow deer and bison all feed on branches, shoots and bark. They in turn are preyed upon by the wolf, puma, tiger (inhabitant of Manchuria) and bear. Although the ruminants and nearly all the large carnivores look for food in the forest glades, they need the concealing protection of ferns, brambles and bushes. Furthermore, in autumn and for part of the winter many animals, from fox, badger and marten to herbivores such as the red deer, turn their attention to fruit growing on low bushes–home of the hazel dormouse (*Muscardinus avellanarius*), the fat dormouse (*Glis glis*) and the garden dormouse (*Eliomys quercinus*). But the most characteristic animals at this level are the birds–warblers, great tits, chiffchaffs, nightingales, blackbirds and thrushes, which feed on hosts of insects and quantities of berries and seeds.

The sparrowhawk nests in shrubs and bushes, well protected here from high wind and storm. Higher up, of course, in the branches of the taller trees, birds are abundant, not needing to

Woodlice are small terrestrial crustaceans which are frequently found in wet regions, including the deciduous forest where they help to decompose dead vegetation.

Facing page (above) : The Virginian opossum is the only marsupial of the Holarctic region and, unlike most mammals of the deciduous forest, is increasing in numbers. It has the habit of carrying the young on its back, is extremely adaptable and has an eclectic diet. (*Below*) The slow-worm is really a legless lizard. This typical woodland resident is hunted because of its resemblance to a viper but is in fact a useful creature, feeding on earthworms, slugs and caterpillars.

The fat dormouse (*above*) is, together with the red squirrel, one of the most typical rodents of the Eurasian deciduous forest. Another forest-dwelling rodent is the hazel dormouse (*below*).

fear reptiles or amphibians. The only mammals at this height are martens, squirrels (the former hunting the latter) and bats, hanging from the branches or concealed in the hollows.

Not all the birds are necessarily found in trees and bushes. Many construct nests in holes abandoned by other forest vertebrates in the ground. The nocturnal birds of prey, very plentiful in this environment, concentrate their attacks on rodents and arthropods hidden in dead stumps and fallen branches. Others hunt the same prey but operate during the day; buzzards, for example, usually nest in the branches, excellent vantage points for surveying the surroundings for victims, but often fly far afield to track prey over open ground. Even more versatile, with a wide range of food, is the goshawk which nests high in a tree yet hunts lizards, mice and hares in the undergrowth as skilfully as it captures magpies, turtle doves, pigeons, blackbirds and woodcocks in mid-air.

Woodpeckers search for insects and larvae in tree bark, tapping away at the vertical surface which is inaccessible for this purpose to any mammals or reptiles. Any small animals which happen to be found on the trunk itself are also liable to be snapped up by these agile climbing birds.

Broadly speaking, the amphibian, reptile and mammal inhabitants of the forest tend to confine themselves to the lower levels, whereas the birds are more opportunistic, being found not only among the higher branches but also taking advantage of any

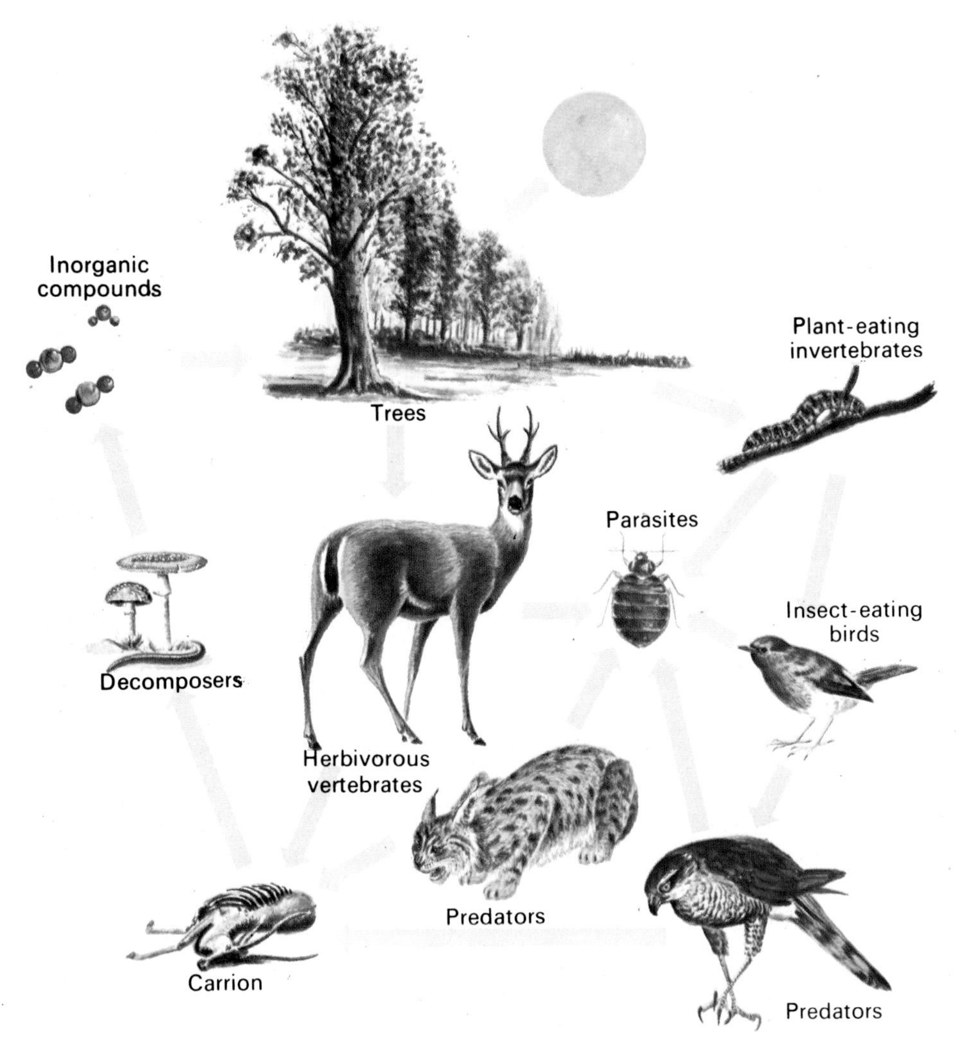

In the deciduous forest, as in other biomes, only the plants (trees, bushes, shrubs and other forms of low-lying vegetation) can transform solar energy, by means of photosynthesis, into energy that can be assimilated by animals. This is then passed on from herbivores to predators and then to decomposers. The last break down organic wastes which are returned to the soil to be utilised by developing plants.

unoccupied clefts and cavities nearer the ground. All these animals play their allotted roles in the forest biome. But whilst every constituent element of a typical ecological pyramid is present–vegetation, herbivore, predator, super-predator and scavenger–it is extremely difficult to place them in rigid categories because there are infinite ways in which the food chains can be arranged. For example, in the insect realm alone there are representatives of all groups from phytophages to carrion eaters. When a scavenging insect is devoured by a lizard which in turn falls victim to a carnivorous mammal, the structure of the food chain becomes enormously complex, not conforming neatly to the comparatively simple pattern which can be traced in biomes where there are fewer protagonists. But undoubtedly the super-predators of the deciduous forests include the wolf, the tiger (in China), the golden eagle and the eagle owl.

The annual round

In spring the forest gradually awakens from its winter torpor. Green buds and dainty blossoms deck the trees as the resident birds bestir themselves once more and are joined by flocks of migrants on their way back from winter quarters. For birds and mammals alike spring is the breeding season. There is bustling activity in the undergrowth and the foliage echoes from dawn to dusk with cheerful birdsong.

This diagram by McCormick shows the scar left by a leaf that has fallen. The leaf comes away from the twig at the base of the petiole after the appearance of a weak area of cells known as the abscission layer. Rain, wind and frost eventually strip the tree bare of foliage.

Facing page : After the long silence and desolation of winter the deciduous forest comes alive in the spring. The snow melts, the streams overflow, the trees are decked with leaves and blossom and the ground is carpeted with grass. For most of the woodland inhabitants this is the breeding season and the forest resounds with the mating calls and songs of birds.

Time passes and the leafy branches of the tallest trees interlock to form a shady canopy of green through which the sun's rays can scarcely penetrate. Summer is at hand. Baby mammals have been born and fledglings have hatched. The main concern of the adult birds is to find sufficient food for the rapidly growing young. As for the mammals, the mothers will probably have given birth in solitude deep among the thickets. They too are preoccupied with the rearing and protection of their offspring, which are avidly hunted by carnivores.

Scant rainfall and heat have combined to dry up streams and rivers, and the few sources of water that remain unaffected are thronged with animals that are unable to survive without drinking at regular intervals.

Already the young animals that are soon to develop into hunters are being instructed in the art of killing. At break of day and at sunset the forest resounds to the hooting of owls. A little later in the year, during September, the first roarings of rutting red deer are heard; and as autumn draws to a close many animals moult, shed skin or take on their winter colours.

Autumn is the season when the forest is rich with its supply of food. Many family ties are broken as adults and young go their independent ways, with more than enough to sustain them. The dry fruits of the oak and the beech are particularly valuable forms of nourishment for a large number of species.

Scientists in the United States made a detailed survey of 2,000 acres of oak forest and calculated that the total yield of edible substances was increasing by 30 tons a year. In Texas it was estimated that a single oak, admittedly a massive specimen, could bear 100,000 acorns annually, although 5,000 was considered an acceptable average yield. In North America red deer (called wapiti or elk), squirrels and mice consume 98 per cent of all fallen acorns, hazelnuts and beechmast.

There is no competition in the forests of the New World from the wild boar but in southern Europe this animal is by far the largest consumer of these food items. A single boar is capable of eating more than 3 lb of acorns at a time, after carefully stripping off the shells. Birds too, especially wood pigeons, are also partial to acorns. The stomach of one wood pigeon was found to contain the equivalent of 32 acorns. In North America starlings, pheasants and wild turkeys are equally fond of the fruits of the oak.

The various parts of the tree–leaves, buds, shoots, flowers and acorns–in fact provide sustenance for more than 200 species of birds. But in an indirect fashion the oak offers nourishment for many other kinds of animals. Hosts of insects feed voraciously on the branches and leaves and these are of course consumed in their turn by countless birds, mammals, reptiles and amphibians. In a sense, therefore, most of the inhabitants of the forest are dependent for survival on the prodigality of this handsome tree.

Apart from the autumn harvest of nuts and acorns, the forest offers a rich variety of pulpy fruits–bilberries, currants, strawberries, blackberries, hawthorn and holly berries, wild plums and cherries. Blackbirds, thrushes and starlings descend

greedily on the bushes and so do normally carnivorous mammals such as foxes, badgers, weasels and martens. Nor is this the end of the food cycle. The seeds of these fruits, dropping to the ground in the excrement of the animals concerned, await the arrival of spring to germinate anew.

Showers of russet and gold

The forest in autumn is truly a place of wonder. Its natural products are many and varied and it throbs to the activities of creatures large and small. But the casual observer may be unaware of all this for the chances are that he will be enraptured with the beauty of his surroundings. The subtle shades of spring and summer green now give way to a glorious range of reds and yellows. The forest blazes with colour, the perennials still resplendent in emerald green, the broad-leaved species garbed in russet and gold, the floor carpeted in brown.

In China, Norther America and even in Europe (though here the hues are rather more modest in range) the deciduous forest in autumn resembles the palette of a painter who has experimentally combined every possible nuance of yellow, ochre, orange and red, dotting the surface with contrasting specks of green and blue.

Leaf fall in autumn is caused by the blockage of the sieve tube by a layer of translucent substance, and the formation of a so-called abscission layer at the base of the petiole (stalk). The cells ringing the petiole become progressively softened until the leaf breaks off. The wound on the stem then heals, leaving a scar. Buffeted by the wind, the tree is soon denuded of leaves and remains dormant for the winter, perhaps concealing a few dormice in its hollows.

There is a scientific explanation too for the marvellous colour changes. The production of chlorophyll, the green pigment in all vegetation, has decreased while the yellow carotenoid pigments, normally concealed by chlorophylls, are now visible. The red and purple leaves of some species are due to anthocyanin pigments, while the reds and oranges are principally created by a combination of substances in the leaf cells.

Scientists want to know more, however, about this entire process. They ask themselves why there is a decrease in chlorophyll production and why the sieve tubes become blocked. They are especially intrigued by the factors which spark off these internal mechanisms as winter approaches, and the way in which the tree stores sufficient energy to survive the cold season and the dry conditions caused by snow and frost.

Many ingenious answers have been proposed but only one seems to have the ring of truth, namely that plants are highly sensitive to fluctuations of temperature and light. It is virtually certain that temperature exerts an influence on the processes of chlorophyll production but many authors believe that much also depends on the amount of light received. In fact the most important factor could be the duration of daylight in any given twenty-four hour period. Experiments in America indicate that the prolongation of the hours of darkness definitely inhibits the production of chlorophyll. Many tests have been conducted to

Facing page : Most of today's deciduous forest are a mixture of broad-leaved and coniferous species. The latter are evergreens so that even in autumn there is a striking contrast with the red, orange and yellow hues of the leaf-shedding trees.

The beech, with its smooth, grey bark and straight trunk, is very common in European forests. It sheds its leaves in autumn and its fruits–beechmast–are consumed in large quantities by wild boars and many other beasts and birds.

verify this theory. Botanists have also subjected an oak to winter-type conditions whilst artificially providing it with the amount of daylight it would naturally receive during the summer. In this simulated situation the tree froze prior to dropping its leaves.

It is claimed, however–and the theory is strongly supported by Soviet scientists–that the annual cycle of a deciduous tree, like the rutting and moulting of animals, is not exclusively determined by its surroundings but equally by an internal physiological mechanism, a type of biological clock or calender. Thus the life cycle of plants and animals is regulated by a rhythm which more or less coincides with the climatic variations of the seasons. This theory appears to be confirmed by the fact that characteristic deciduous species, transplanted to Mediterranean latitudes and then subjected to stable weather conditions, also lose their leaves in the normal manner, although in a somewhat random, disordered pattern.

Unfortunately, jet aircraft, motor cars and factories have so polluted the atmosphere that, according to information provided to delegates of the World Wildlife Fund Congress of 1970, the amount of solar energy reaching the earth's crust is noticeably less than it was several years previously. If this undesirable trend continues, it could alter the biological structure of living organisms, with unpredictable consequences, possibly culminating in disaster on a world scale.

Winter silence and desolation

Hardly have the three summer months passed than the leaves begin to drop, later carpeting the forest undergrowth with gold. At the same time most of the local animal inhabitants prepare for a departure to warmer climes or for a long winter sleep. Although some of the hardier species defy the elements with an appropriate change of pelage or plumage, most of them spend a greater proportion of the day in shelter.

When the branches are covered with a thick layer of frozen snow the problem of securing food becomes more difficult for species such as the wild boar and the roe deer; and the resident grouse, partridges and thrushes now have to seek new places of refuge. It is during the winter that the many perennials of the European forest come to the rescue, offering food as well as shelter. In most deciduous forest areas of Asia and North America, the animals that do not migrate increase their consumption of branches, roots, seeds, young plants and bark. But would this suffice if there were no conifers?

Dr Castroviejo has demonstrated that in the deciduous forests of the Cantabrian mountain chain, where there are no pines, the most important species for local animals is the common holly (*Ilex aquifolium*). Although not normally a large tree it may here grow to 40–50 feet. The holly's stiff, tough, prickly leaves often form a kind of protective roof in winter. After a heavy fall of snow the temperature inside a holly bush is several degrees higher than it is outside and consequently provides a welcome refuge for many tree-dwelling birds.

Winter of course has its incidental joys and both the fir and the holly play their traditional decorative roles during the festive Christmas season. But as far as wildlife is concerned it is undoubtedly the holly which fulfils the more valuable purpose. It is no exaggeration to say that many animals would perish were it not for this hardy shrub with its small red berries. Dr Castroviejo has stated that in the Cantabrian mountain region the only active vertebrates during the winter are:

1) Phytophages (principally rodents) which by reason of their habits and size are able to find their food in the lowest vegetational layer; two bird species, the partridge (*Perdix perdix*) and the red-legged partridge (*Alectoris rufa*), which eat the same type of food, hunting for it at the edges of streams where snow does not settle; and other phytophages that feed mainly on holly.

2) Insectivores which feed chiefly on arthropods and other invertebrates in a lethargic condition.

3) Predators: hunting animals belonging to either of the preceding groups.

In exceptionally bad weather, when the forest is swept by snowstorms and high winds, red deer, roe deer, wild boars, squirrels, woodcocks, thrushes and other species sometimes spend weeks on end sheltering in or under holly bushes, with all necessary food to hand. Only if hard pressed–possibly by an enemy–will they abandon these warm thickets. The hunters of course are especially active in winter and the tracks of both predators and prey are clearly etched in the snow.

The days gradually lengthen, the snow begins to melt and the rivulets swell into rushing torrents. In the course of the winter

Winter cold, lack of water and shortage of food compel many forest animals either to depart for warmer climes or to hibernate. Deep silence reigns in the undergrowth as plants lie dormant under a covering of snow. The only noise is that of the bare branches waving or snapping off in the wind.

a large number of animals will inevitably have succumbed to cold and hunger. Now the survivors can once again prepare to make good their losses. The sun's rays give out a little more warmth with every passing day and the first tentative twittering of birds announces the approach of spring.

The exotic forests of the Orient

The climate of the Chinese seaboard, similar to that of Korea and the islands of Japan, is influenced by the warm currents of the Pacific Ocean and there is a gradual transformation of scenery as one travels from the heart of this enormous land in the direction of the sea. The coniferous belt of taiga gives way to stretches of mixed woodland and then to the deciduous forest proper; and as one journeys southwards this yields in turn to a forest belt of tough perennial species, similar to laurels. Finally, along the coast of southern China and the shores of the Gulf of Tonkin, as well as on the island of Taiwan (Formosa), there are areas of subtropical and tropical woodland which have their European counterparts in the forest regions of the countries of the western Mediterranean.

With such a rich diversity of woodland and forest habitats it comes as no surprise to find an equivalent wealth of wildlife. Animals which are normally regarded as belonging to entirely different biomes are here discovered living together. Unbelievably, roaming these vast Manchurian forests alongside typical carnivores of the Palearctic region such as foxes, weasels, martens, bears and lynxes, are the largest tigers in the world, though recent estimates unfortunately put their number at no more than about one hundred. Other forest hunters of eastern Asia include the panther, the leopard cat (*Felix bengalensis*), the Himalayan black bear (*Selenarctos thibetanus*) and the strange raccoon-like dog (*Nyctereutes procyonoides*)–an ancient inhabitant of these forests which has spread into central Europe.

These predators hunt a wide range of familiar animals, including red deer, roe deer, elks, squirrels, mice and shrews, as well as localised species such as the goral (*Nemorhaedus goral*)–a small Himalayan ungulate–the musk deer (*Moschus moschiferus*), the sika or sika deer (*Cervus nippon*), the short-eared Chinese hare (*Lepus sinensis*) and other small mammals.

Birdlife is equally remarkable. Mingling with all the characteristic Oriental species are birds adapted to the various habitats encircling this deciduous forest belt–pheasants and grouse alongside golden orioles, blackbirds and other woodland species; and, on the plains of the river valleys, a mixed population of bustards, partridges, Japanese cranes (*Grus japonensis*), pied harriers (*Circus melanoleucus*) and mandarin ducks (*Aix galericulata*).

The animals of the mixed forests of the Far East originate in Siberia, the Himalayas and the tropics; yet here they mingle freely with species common to the Eurasian steppes and the woods of Europe and North America to make up one enormous, integrated community which from the naturalist's point of view is one of the most important and fascinating in the world.

The raccoon-like dog originated in the forests of eastern Asia, was introduced into Russia and made its way westward into Europe. In appearance it is a cross between a fox and a badger. Most active at dusk and by night, it sometimes ventures out during the day to hunt birds, insects and small mammals at ground level.

Facing page : The forests of Japan, with their magnificent variety of trees, harbour many interesting animals. Flora and fauna attract naturalists from all over the world.

CHAPTER 8

The wild boar: savage forest omnivore

Europe has never been renowned for an abundance of large predatory mammals and the few that once roamed the forests have now almost completely vanished. The wolf and the lynx, it is true, still roam the more sparsely populated regions, as do a host of smaller carnivores. The brown bear is much rarer and relatively peaceful by comparison. All these animals are certainly wild in the sense of living free in their native habitats yet in the popular mind they do not rank with the Big Cats and other hunters of tropical countries.

There is nevertheless one woodland animal remaining from primitive times which is still capable of invoking feelings of genuine awe. In the depths of the deciduous forest (though long extinct in the British Isles) dwells a large mammal which seems to be a relic of the age of the mammoth and the woolly rhinoceros, a creature that is fearless, savage and dangerous. This is the wild boar (*Sus scrofa*).

The head and body of this sturdy animal may be likened to two cones, joined at the base and flattened at the sides–no mere caprice of nature but an evolutionary adaptation to a forest environment, enabling it to move with ease and at considerable speed through the thickest undergrowth. The very long muzzle ends in a sensitive, mobile snout with large nostrils which is used not only for raking over dead leaves but also for digging in the earth for roots, bulbs, fungi and small animals.

The wild boar has little in common with its farmyard relative apart from a habit of wallowing in mud. Most people draw the conclusion that all pigs are therefore dirty animals, yet this is a mistaken impression. For the wild boar a mud bath is obviously a pleasurable activity but it is much more than a pastime. This is the way in which the animal rids its pelage of parasites and

Facing page: The wild boar still enjoys an enormous distribution range, inhabiting many forests in Europe, Asia and North Africa. Because the population is so scattered there are striking variations in size and behaviour, with naturalists distinguishing several subspecies.

Geographical distribution of the wild boar.

WILD BOAR
(Sus scrofa)

Class: Mammalia
Order: Artioactyla
Family: Suidae
Length of head and body: up to 71 inches (180 cm)
Length of tail: 10–18½ inches (25–47 cm)
Height to shoulder: up to 40 inches (100 cm)
Weight: up to 750 lb (350 kg)
Diet: omnivorous
Gestation: 16–20 weeks
Number of young: 2–4, 6–8, 8–12, depending on age
Longevity: 10–12 years in wild, 20 years in captivity

Adults
Massive body, triangular head prolonged by snout. The pelage, consisting of thick bristles and some finer hair, grey, blackish or brown (generally silver-grey in summer and dark in winter). A stiff mane extends from the head down the back. Winter pelage appears between October and May. The feet and the surrounds of the snout, which is grey and naked, are always black.

Young
The piglet is born with a short snout and striped pelage consisting of dark longitudinal bands on a reddish ground. Within six months the hair has turned uniformly reddish and at a year is the same colour as that of the adult.

Facing page : It is not generally appreciated that the wild boar is an exceptionally clean animal, frequently immersing itself in water and wallowing in mud. The apparently dirty mud-bathing ritual has a hygienic function, ridding the body of parasites and keeping the skin in sound condition. These sites are carefully chosen in areas rich in mineral salts.

keeps its skin clean and healthy, thanks to the mineral salts and organic substances contained in the slime.

We have already come across similar examples of what might be termed therapeutic mud-bathing in the animal world, buffaloes, hyenas, elephants and rhinoceroses being just a few that indulge in it for hygienic reasons; and what animals discovered by instinct we have lately confirmed scientifically. Mud is unquestionably good for the skin as women undergoing beauty care and health treatment will readily testify!

It is interesting to note that a wild boar, like an elephant, will not simply lie down to roll in the puddle nearest to hand but will carefully select its wallow, journeying if necessary several miles, even at night, to find a convenient site. Somehow it manages to choose mud of precisely the right quality and composition for its purposes.

An ingenious hunter

What is the secret of the wild boar's ability to survive in such a variety of forest environments, extending from western Europe to the shores of the China Sea ? One of the principal reasons is an enormously diverse range of food. It will in fact consume almost anything it can find in the undergrowth. As far as vegetation is concerned it shows a special partiality for chestnuts, acorns and all kinds of nuts (using its apparently clumsy lips and teeth to remove the husks) and for edible fungi, including truffles. Its carnivorous tendencies have not been so fully documented. Animal protein is furnished by rodents, rabbits and hares, reptiles, arthropods and other invertebrates, and even carrion. Furthermore, in summer it frequently goes fishing, trapping freshwater species sunning themselves near the surface as well as amphibians and molluscs.

Some wild boars exhibit a high degree of specialisation. Professor J. Lalanda once spent several weeks following the tracks of a huge boar which hunted rabbits to the exclusion of all other animals, showing uncommon strength, ferocity and ingenuity in dislodging victims from their burrows. It methodically blocked up all possible exit holes apart from one, using this last as a subterranean excavation route and trapping one animal after another in the side galleries.

Comparative surveys have led scientists to the conclusion that individuals which hunt other animals are better developed physically than those preferring a predominantly vegetarian diet.

In some regions the wild boar eats numbers of vipers and scorpions, so much so that there is almost certainly an inverse relationship between the population densities of the mammals and the venomous reptiles.

The rooting habit of the wild boar–using its flexible snout to root out edible substances from below ground–is undoubtedly of considerable importance in the forest's ecology. This busy digging activity aerates the soil in much the same way as does a plough, creating conditions favouring the germination and development of new plants. Consequently the mammal performs a positive role in stimulating forest growth.

After a mud bath a wild boar will rub its flanks against a tree trunk in order to fortify its thick layers of hide. It also scratches the bark with its tusks. The marks left on the tree warn other boars that this particular mud puddle is occupied.

The shoulders and flanks of a wild boar are covered by an extra-thick layer of skin, helping to blunt the teeth thrusts of a rival in fights during the breeding season.

Weapons and armour

The four canine teeth of the wild boar constitute its principal weapons and are kept in prime condition throughout life. The lower pair are the tusks, up to a foot long, curving slightly to the rear, with roots extending back in the gums to a point below the molars. They hone the lower canines. The upper canines are smaller and grow straight down and slightly outwards. The four teeth are so arranged that the front face of the upper pair is in constant contact with the rear face of the lower pair, so that they rub against each other whenever the jaws move. Seen in profile the tusks resemble miniature scimitars and are as dangerous as they look. An attacking boar will charge, mouth agape, at the enemy, attempting to rip the flesh with a powerful downward head motion. Should the tusks find their mark the resultant wounds are likely to be serious, even fatal. Hunting dogs have had their tails sliced through at the root with surgical-type precision. The teeth are of course equally effective as cutting and raking tools. Those of the sow are very much less developed.

During the breeding season the boars engage one another in violent fights for possession of the sows. But these aggressive animals, capable of disembowelling an enemy, seldom injure others of their kind. Like most other animals equipped with powerful natural weapons, they have evolved ways and means of minimising damage when fighting among themselves. In their case it depends on anatomy more than behaviour. All adult boars have an extremely tough layer of skin on shoulders and flanks so that the body is virtually armour-plated. As two rivals circle for an opening they first position themselves side by side, flanks touching. Then each boar tries to sink its tusks in the other's neck, this being the most vulnerable part of the anatomy, but it is usual for the canines to lodge in the opponent's thick hide, thus automatically blunting the blow. Large males killed during a hunt often display long scars on their sides but these are little more than skin deep.

Incidentally, wild boars are not alone in possessing this thick protective skin. The North American Indians used to ward off the bullets of the first white colonists by covering the exposed parts of the body with strips of buffalo hide.

After the boar has completed its mud-bathing ritual it trots over to a nearby tree and rubs its body vigorously against the rough bark. Then it scratches the trunk with its teeth. This exercise appears to have a double purpose. The site is clearly marked so that potential rivals can be in no doubt that it is private territory; and the rubbing activity hardens the skin after the softening effects of the liquid mud.

There is another aspect of the behaviour of this species which is evidently related to the previous activity and which needs further investigation. In southern Andalusia and in Morocco hunters have seen the piglets approach their father while he is taking a rest and then slap their snouts smartly against his flanks. The latter shows no signs of irritation and makes no effort to shake the youngsters off, suggesting that this too may be a form of hide-toughening massage.

The sow on the left may be distinguished from the male by her narrower snout and smaller teeth. Neither sex is naturally aggressive unless attacked but the sow is especially dangerous when accompanied by her piglets and will charge at intruders.

Attack and retaliation

Much has been said and written about the aggressive nature of the wild boar and it is no simple matter to sort out the truth from the tangle of hasty judgments based on third-hand report or sheer ignorance. There is no denying the fact that this formidable animal shows little hesitation in facing foes or rivals of its own species, other forest animals, and even its worst enemy, man. Strength, speed and astonishing agility are reinforced by the razor-sharp teeth so that its fighting prowess is not in question. Nevertheless, experience proves that the wild boar chooses to avoid a confrontation whenever possible. Only when it is trapped in a tight corner, with escape routes cut off, will it turn to challenge its adversary; and at that point it is transformed into a really dangerous animal.

The sow may be just as fierce as her mate, especially when rearing her litter. It is the height of rashness to try venturing too close at such a time for the mother will charge in blind rage at any intruder. Her teeth are shorter than those of the boar but still sharp enough to inflict deep gashes, and a kick from her flailing feet can also send the enemy reeling. In such situations it is surely no exaggeration to claim that the female is more deadly than the male.

The extraordinary courage of the wild boar is evident from the manner in which it confronts wolves. The two species are implacable enemies. A forest warden once described how he was able to reconstruct the pattern of one such battle by marks left in the snow. Five adult wolves had apparently attacked a solitary boar and succeeded in killing him, but there were five separate bloodstained tracks leading away from the spot where the carcase

Following pages : The piglets of the wild boar are very precocious, capable of standing upright only a few hours after birth. For the time being, however, they huddle against one another for warmth and spend their time asleep or suckling. After about two weeks they are out rooting for vegetation and insects. The distinctive striped pelage–an effective camouflage–vanishes after six months and the hair turns progressively darker as the animal gets older.

Solitary males often clash in the rutting season. At first the rivals rub flanks (1) and make no attempt to strike a blow. Then one of the boars will suddenly launch an attack (2), trying to bite the other's neck. This apparently violent duel is really part of a recognised ritual for the tough hide on the flanks usually averts serious injury.

lay, indicating that all the predators concerned had sustained injuries in the course of the fighting.

We should bear in mind that mammals, during their long evolution, have acquired a certain flexibility of behaviour which enables them to adapt successfully to new situations. In this respect they differ from insects and other zoological groups with more rigid behaviour patterns. In many mammals, therefore, aggression is a variable factor and will depend much on the personality and experience of the individual. The wild boar is no exception. Some of them possess unusually long legs and evidently react instinctively to danger by fleeing; others, smaller in size and with shorter legs, can be counted on to face their enemies with boldness and determination.

Paradoxically, young boars raised in captivity tend to be relatively docile and often grow up to behave in the same way as dogs, sticking close to their master and even forming part of a hunting pack when the quarry is itself a wild boar. Although there are few shackles to their activities no cases have been reported of such domesticated individuals returning to the wild.

Boars, sows and piglets

It is difficult to generalise about the habits of the wild boar. Over the centuries the animal has been hunted with such fury and intensity while its traditional habitats have been destroyed or modified to an extent that there is hardly any place left in Europe where it can be said to live under genuinely wild conditions. But although study of the animal is difficult it is evident that it is continually on the move, no sooner settling in one district than it is off to a new destination. Such journeys do not seem to be associated with the quest for food, being caused principally by human activities. The wild boar abandons its refuge because it has been disturbed, moving off in search of more tranquil surroundings.

Long experience has probably taught the wild boar that of all enemies man is most to be shunned and for this reason it is mistrustful by nature. It will, for example, constantly change its itinerary, seldom adhering to fixed paths. In this connection it is interesting to consider the selective influence–usually a negative one–that man has unconsciously brought to bear on the evolutionary development of certain animals. The object of big game hunting and associated diversions is normally to procure the most impressive trophy possible–whether it be a finely coloured and patterned skin or a magnificent pair of antlers. Thus the coveted victim will frequently be a large, powerful male. Under prolonged hunting pressure it has been shown without any doubt that there is a gradual deterioration in the physical attributes of the species involved–a slight decline in strength, weight and presence in each succeeding generation. The reason for this is that as the best endowed males are picked off by the hunters there is an increasing proportion of less virile reproductives. Should the latter have even slight defects these are likely to be inherited and passed on by their offspring, with potentially grave effects for the community.

In the course of his wooing ritual the male circles his partner and gives her affectionate taps with his snout. He also lets out rhythmic grunts which have the effect of bringing her to an immediate standstill.

The rutting season usually extends over the second half of December though it may be prolonged until the end of January. Some animals have been seen attempting to mate at more or less any season but it would appear that such individuals are products of cross-breeding with domestic pigs–common in some regions.

Breeding time sees the older reproductives journeying long distances in quest of partners, sometimes neglecting to eat properly and growing quite emaciated as a result. The females are often found in the company of young mature boars which take off as soon as they catch sight of the approaching solitary male. But if a sow has already been commandeered by a powerfully developed boar the dissuasive methods of the scouting male may be of little avail and the rivals will come to blows. An exceptionally vigorous boar may be able to win the favours of several sows; harems of six to eight females are not unknown.

Once the threat of competition has receded the dominant boar turns his attentions to his mate and begins wooing her in a rather clumsy fashion, she accepting his advances without any signs of fear or annoyance. Having brought her to a standstill he circles round, nudging her with his snout and urinating frequently. The little bites which he occasionally plants in her back are not intentionally aggressive but merely somewhat over-ardent demonstrations of tenderness. The female reciprocates and both animals rub muzzles affectionately.

During this whole courtship ritual the male lets out rhythmic grunts which seem to have the immediate effect of immobilising the female. J. F. Signoret and other French scientists studying the behaviour of domestic pigs have remarked that the cries emitted by the males during the breeding season have precisely the same result. When the naturalists played tapes on which these sounds had been prerecorded to young sows that had never mated, 71 per cent of them remained rooted to the spot and even allowed the scientists to sit on their backs. Later experiments had similar effects. The conclusion must be that an instinctive mechanism is involved since there is no question of such behaviour having been acquired by previous experience.

The sows give birth after a gestation of sixteen to twenty weeks, either in April or May. Those with their first litter generally have two to four piglets, others normally give birth to between six and eight, though the number may be anything up to twelve. Litters of sixteen have been recorded but it is probable that some of these were the offspring of another mother, perhaps adopted orphans, as is a common practice. The sow will have prepared a rudimentary nest for the litter, about 12–16 inches high and 3 feet in diameter, well concealed under a mound of grass, dry ferns, moss, heather, twigs and other scraps of vegetation. It is always situated in a fairly inaccessible part of the undergrowth and to provide added protection some sows top it with a roof-like network of large branches.

The newborn piglets have very short snouts and are covered with soft reddish hair. Alternating light and dark bands down the back and flanks have a remarkable camouflaging effect. Four of the ten teeth with which they are born are canines. Their eyes are open and they can stand after a few hours, although they do

To give birth and rear her litter the sow digs a kind of nest some 3 feet across which she conceals under a mound of leaves, twigs, roots and other vegetation.

The baby wild boar is characterised by a short snout and alternating light and dark brown bands along the length of the back. At this stage the hair is silky.

Facing page : While the older males usually live a solitary existance, wandering at random through the forest, the young boars and adults of under five years of age are much more sociable, living in small groups.

not leave the nest for some days, spending their time sleeping and suckling, huddled against one another for warmth. The mother wanders off from time to time, first covering them with vegetation. Only after a week or so do they begin following her and soon they do not even bother to return to the nest at night. During their first fortnight they scatter at the least hint of danger, then stand absolutely still to avoid detection.

The piglets are suckled for about two and a half months but within two or three weeks of birth they are already rooting in the earth, feeding on vegetation and small animals. The sow suckles her youngsters by settling on her side and emits low grunts in the process. Since not all the milk-producing channels are operating effectively immediately after the birth, there may be violent fights and casualties among the piglets as they manoeuvre for the best positions. But in due course some sort of order is established, with each piglet allotted its appropriate teat.

The younger her offspring the more aggressive the sow tends to be towards intruders. The only adults of her own kind that are tolerated are other mothers and together they form large herds that may include animals of up to three years of age. Only when the breeding season approaches will the females abandon this group and as soon as the progeny of the new litters are capable of standing the mothers will once more come together.

At the age of about six months the young boars have light brown hair, the stripes have faded and by then their dentition is complete. A year later the hair has turned dark brown or black and the long tusks of the males are clearly visible. By the time they are two and a half years old they have left the family circle and stray off in groups. At the age of four or five years the males have adopted a solitary life.

Some wild boars are already sexually mature at eight to ten months and it has been claimed that if food is plentiful about 50 per cent of females under a year old are capable of mating. Normally this figure is nearer 10 per cent and in years when the weather is very bad none of these young animals will be on heat.

Under exceptionally favourable conditions a sow may have two litters a year, one in spring, the other at the end of summer. In some regions, therefore, the wild boar population may increase dramatically, more than compensating for losses incurred as a result of hunting or epizootic diseases.

Sense and intelligence

Contrary to what one might think, the wild boar is a relatively intelligent animal. Although its sight is poor, the senses of hearing and smell are extremely well developed. Yet that is not the whole story. Many hunters have remarked that in certain situations in which neither sound nor smell are involved, the animal is still capable of sensing the presence of an intruder. R. Hainard, in his book on European mammals, attributes this phenomenon to a kind of sixth sense which enables it to detect ground vibrations and thus follow unusual activities occurring beyond the normal range of perception.

The wild boar's wide food choice testifies to its adaptability.

It has a way of solving problems with a remarkable sense of deliberation. When cereal crops are too high for the animal to reach the ears, instead of breaking the stalks off one by one it tramples a small circular patch and scoops them up in bunches. Moreover, it takes advantage of the work done by farmers. At harvest time, for example, it slips into a field under cover of darkness and then, taking not the least notice of any corn still lying on the ground, makes for a sack of grain, rips it open with its tusks and devours the contents.

An older boar learns much from long experience and resorts to all manner of ruses to avoid traps set for it. One astute trick sometimes used when surrounded by beaters is to get a younger animal to show itself and act as a decoy.

Wild boars communicate with one another by means of body movements and sounds. There are a number of recognised vocal signals. A long snuffling noise expresses anxiety and may serve as an overt warning, while a deep grunt is an alarm call, inviting all within hearing to take to their heels without delay. The sow has her own low grunt for summoning her young and the latter respond by grinding their teeth.

It is by studying the behaviour of the piglets that one appreciates how widely individual personalities may differ. Some are clearly very timid and remain close to their mother's side, while others have no hesitation in going off to play by themselves. Then there are those that show extreme independence, taking advantage of their mother's momentary lapses of attention to wander away on exploratory journeys. These precocious youngsters cause all kinds of problems for the mother. The need to keep constant watch on them or go searching for them when they are lost may place both herself and the other members of the litter in a highly dangerous position.

The thick, dark winter coat of the wild boar, apparent from October till May, protects the animal from the extreme cold in the more northerly parts of its range.

The wild boar in Spain

The wild boar is the only large wild mammal which, without having been reintroduced, is still found in reasonable numbers in many mountainous forest regions of Spain. These are the most southerly refuges of the European population. The animals are smaller than those inhabiting the northern part of the continent, a phenomenon which is not unique to the species since hares, bears and deer of northern Europe, to mention only a few, are markedly heavier than their southern counterparts in countries that border the Mediterranean.

The explanation of this size and weight discrepancy has something to do with animal adaptation to the colder climates of these northern regions but it is also the consequence of a much richer range of food. To take deer as an example, it is immediately apparent that Mediterranean subjects imported from central and eastern Europe (animals characterised by their weight and magnificent antlers) lose these attributes after only a few years of life in the south. By that time they will be no larger, and may in fact be somewhat smaller than deer born and raised in these parts, for the latter, in the course of centuries, will have adapted to the difficulties of such arid climes.

Facing page: The wild boar is more than ever active in winter when it is compelled to make long journeys and to spend much time rooting for suitable food. It is adept in scooping out worms, larvae and roots from the frozen soil.

Facing page : Few large animals are abroad in winter but the wild boar indefatigably explores the tips of the undergrowth now covered by several feet of snow.

This holds true for the wild boars of the Iberian peninsula. In comparison with their enormous 480-lb relatives of the German forests, not to mention the gigantic animals found in the Carpathians which weigh over 750 lb, the largest solitary males in Spain seem ridiculously puny at 200–250 lb.

Despite fairly intensive hunting activity in this region, the wild boar population has prospered in recent years. This increase is partly due to the disappearance of the wolf, their principal predator, and partly to reafforestation which has transformed bare hills and mountainsides into impenetrable thickets. Sheltered by dense plantations of young pines, inaccessible to the public or to domestic herds which might damage the developing trees, the wild boars are able to breed in virtually undisturbed conditions. The size of the population is already causing disquiet for it is realised that if nothing is done the animals will become a plague. Here is yet another example of natural imbalance. Artificial elimination of competing species always brings repercussions and usually leads to trouble more serious than had been experienced in the first place.

What can be done to check this dangerous trend? It is unthinkable that these useful forest animals should be exterminated, so that the only solution must be to limit the extent of the damage they may cause. It is known, for example, that they are very sensitive to certain odours. Petrol apparently keeps them at a respectful distance. Might it not be possible to make use of this established fact to ring these regions with specially treated scent posts? It is in fact surprising that in view of the failure of other measures some such experiment has not been attempted.

The tell-tale prints of a wild boar in the soft snow are only too easily recognised by hunters for whom there is no close season.

The wild boar population of the Iberian peninsula is fairly well distributed. Their tracks have been found at alpine level in the Pyrenees where they have laid waste the soil and denuded it of every vestige of plant life. It is more surprising still to discover them among the stunted oaks of central Castile, areas surrounded on all sides by cultivated land. Here the animals systematically ravage fields of corn and potatoes at night. They are also numerous among the sand dunes and marshes of the Guadalquivir basin at sea level, principally in the Doñana reserve where they are concealed by gorse and sedge.

In regions where they are not hunted the animals display a great measure of confidence, using their sense and intelligence to avert danger. The older boars are particularly sure of their strength, venturing out in broad daylight, facing enemies with extraordinary courage and even putting large dogs to flight, sometimes after inflicting serious wounds with their tusks. On the other hand, where hunting is prevalent, the boars are extremely reticent, seldom entering inhabited localities, spending most of the day under cover and roaming open ground only at night when all suspicious noises have ceased. Even then they sniff the air and listen carefully before venturing out by moonlight.

Choice of food will vary considerably according to place and season. Jesús Garzón Heydt has stated that in very cold weather the wild boar roots more frequently than at other times, covering large stretches of ground in its search for earthworms, insect

The heads of the boar and sow are very different. The muzzle of the boar (1) is elongated and the well developed canines are so positioned that the upper and lower pair rub against each other whenever the jaws move. The sow (2) is much smaller, her snout is shorter and her teeth are not nearly such powerful weapons.

larvae, bulbs, fern roots, chestnuts, fir-cones, buried acorns and similar items. In spring and early summer it spends hours nibbling succulent new grass; and with the arrival of warmer weather its tastes become increasingly carnivorous. It hunts interminably for ants, consuming the larvae, the nymphs and the food reserves collected by the workers. If need be it overturns large stones in order to dislodge scorpions and any insects or small animals concealed underneath. The nests of fieldmice are methodically destroyed, both young and adults being the victims, and all the food gathered by the rodents and lying in special chambers is likewise devoured.

Boars perform a valuable service in pine forests by feeding on harmful insects, especially processionary caterpillars which in summer are about to turn into butterflies. But they are undoubtedly a nuisance to farmers and fruit-growers, causing extensive damage to crops of wheat and other cereals as well as to vineyards, fig plantations and olive groves. Their predilection for chestnuts and acorns rouses less controversy.

The long muzzle and snout of the wild boar form a valuable tool which helps it to root in the hardest soil.

With so much food available at this time of year it is not surprising that the animals look uncommonly fat and heavy. But the breeding season is at hand and the males soon begin to neglect their food. By January, after mating has occurred, they are extremely thin and many of them are further weakened by the wounds they have sustained in intraspecific fighting.

The animals feed on carrion all the year round and although actual contaminated subjects are few and far between, this habit does make them potential carriers of trichinosis, a disease which may be transmitted to humans eating pork products infected by a parasite, the adult of which lives in the animals' intestines. The only safe course is to cook wild boar meat very thoroughly.

The behaviour pattern varies from season to season and also depends on the age of the individual concerned. Sows that have recently had a litter, for example, show a particularly strong territorial instinct, hardly ever moving from the parts of the tangled undergrowth where they have concealed their young. As the piglets grow the whole family wanders off to look for food and to root with tremendous vigour.

Thus the wild sows of the highlands of Estremadura, which normally select the warm, sheltered valleys of Caceres to give birth to their young, tend to stray off when summer arrives towards the cooler slopes of Salamanca, conveniently situated close to the cereal-growing areas where they can feed at leisure on the ripe ears of corn.

By rubbing against tree trunks the wild boar leaves characteristic marks, often scraping away the bark. It may also plough up the surrounding ground.

In late summer and autumn the sows return to the districts where the corn is now ripe and the fruit trees heavily laden. As the chestnuts and acorns tumble to the ground whole herds converge on woods and forests, deserting the areas where until quite recently they had been seen in large numbers. The solitary boars, on the other hand, show no inclination to leave the regions with which they are so familiar. Normally they are content to remain in one place all the year round.

From time to time, however, some of the older boars venture on long journeys, the purpose of which is not known. Their movements appear to be quite arbitrary. One day they may be sighted in a locality which appears to suit them perfectly and the next day they are off on their travels, for no apparent reason. They cross hills and plains before settling down in a district which may be dozens of miles from their previous home and which does not seem to offer any special facilities that they did not enjoy before they started the expedition.

It may not be fanciful to suggest that these animals, which are powerful and intelligent enough to withstand all the attacks of their enemies, are at such times in the grip of the same kind of irresistible urge which drives some people to give up a comfortable existence and go out to seek new horizons, even at risk to their lives.

CHAPTER 9

Red and roe deer: lords of wood and forest

Civilised man can no longer claim that his survival depends on his success in tracking and killing wild animals. Yet there is some kind of primitive urge which makes it impossible for him to renounce an activity which has little social value or moral justification; and by virtue of modern developments of firearms and ease of transport, hunting, whether for sport or profit, is bound to have wide-reaching repercussions not only on the species directly concerned but also on others closely dependent on them. Such animals fall broadly into two categories–those herbivores that have traditionally been regarded as game and those carnivores that prey on such animals in the wild in order to survive. The predators are blandly assumed to be harmful (to man's interests) and this has often been taken as sufficient justification for persecuting them. Consequently some species have been brought to the verge of extinction. Thanks to the efforts of conservationists, increasingly backed by public opinion, many hoofed animals have for the moment been spared.

As a direct result of the intensive campaigns against the predators, many animals now freed from the marauding activities of traditional enemies prospered at their expense. In some cases the population explosions proved costly and destructive, yet the trend was accelerated by the enactment of laws and decrees–enforced by governments, local authorities and private individuals–designed to restrict hunting. Where certain favourite game animals were concerned this played into the hands of vested hunting interests for there were many more available for shooting. Principally involved were those most coveted of all victims, the various members of the deer family, and none more so than the red deer (*Cervus elaphus*).

The red deer and related species are characteristic inhabitants

Facing page: The red deer and its close relatives are widely distributed throughout the deciduous forest regions of Europe, Asia and North America. The stag, with his magnificent horns, is a noble animal and probably the most coveted prize of hunters in the Holarctic region. Yet his appearance belies his behaviour for he is a timid beast which flees danger without giving any heed to his companions.

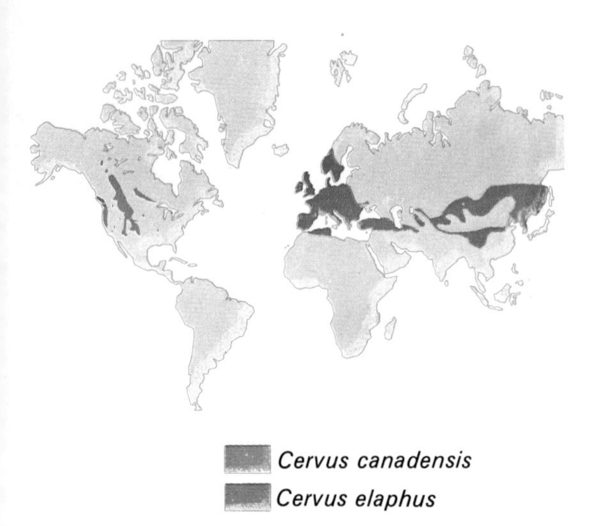

Cervus canadensis
Cervus elaphus

Geographical distribution of the red deer (*Cervus elaphus*) and the wapiti (*Cervus canadensis*).

RED DEER
(*Cervus elaphus*)

Class: Mammalia
Order: Artiodactyla
Family: Cervidae
Length of head and body: up to 100 inches (250 cm)
Length of tail: 4¾–6 inches (12–15 cm)
Height to shoulder: 48–60 inches (120–150 cm)
Weight: male 220–550 lb (100–250 kg)
female 175–265 lb (80–120 kg)
Diet: herbivorous
Gestation: 225–270 days
Number of young: 1, rarely 2
Longevity: 15–20 years

Adults
The stag has a powerful chest, plump belly, sturdy neck and long, slender legs. The hind is smaller and altogether more graceful. Summer pelage is reddish-brown, the tail white bordered with black; the winter colour is greyish-brown, often very dark, especially on the neck. Only the male has antlers, which are dropped annually in March and have regrown by the beginning of July when the velvet is also shed (though the date may vary according to the individual).

Young
Reddish-brown coat with white-dappled neck and back.

Facing page : A red stag in rut may collect a harem consisting of a number of hinds, depending on his strength and ability to repel rivals. Despite his presence the hinds maintain their normal matriarchal herd structure and it is invariably a female that controls the movements of the group during the breeding season.

of the deciduous forest regions of Eurasia and North America. They are also found in the Mediterranean thickets and–as a result of deforestation and the spread of agriculture–on arid hill and mountain slopes.

There is a wide range of distribution with equivalent variations in size and weight. Thus the Barbary red deer weighs a little more than 200 lb while the subspecies of the northern tundra is over 750 lb. On the other side of the Bering Strait there is an even more massive deer, the American wapiti, held by local zoologists to belong to a separate species but according to Soviet scientists descended from the Siberian race. Also in the New World there is the very abundant white-tailed or Virginian deer, and, threatening to overspill its somewhat restricted habitat in the western states of the U.S.A., the mule deer.

Because these deer are so plentiful in most deciduous forest regions naturalists have been able to study their behaviour in great detail and there is an impressive quantity of information on the subject. One of the most eminent men in this field is the British naturalist Sir Frank Fraser-Darling, whose book, *Natural History in the Highlands and Islands*, following years of research in the Scottish Highlands, is a classic. The information in these pages is to a large extent based on his personal observations, supplemented by facts supplied by naturalists from other countries.

The social life of the red deer

Red deer are gregarious animals which tend to live in separate herds: the females or hinds which, accompanied by their young of varying ages, herd together; and the males or stags, aged three years and over, which form their own quite separate herds. These two social groups go their different ways for the greater part of the year and are thus distinguished not only by appearance but also by contrasting behaviour patterns.

The female herds usually consist of family units, each made up of an adult hind and two or three calves. There is a strictly observed social order governing these little family groups which comes into operation whenever they come together. But although each family occupies a discrete section of the home range there is not normally any lasting contact among them.

At different seasons, especially in fine weather the herd breaks up, family units forming temporary associations and wandering off in different directions, usually guided by the oldest, most experienced hind. At other times all the hinds occupying the same ground will intermingle and it is immediately evident from their attitudes which of them is in control, invariably a mature individual with at least one calf who, while all her companions graze and quietly ruminate, displays signs of nervousness and appears to be constantly on the alert.

In everything they do the hinds display a greater degree of purpose and cohesion than the stags. For one thing the bands of males are much more loosely organised, being in fact no more than random aggregations of individuals, each of which enjoys complete independence. If there is a leader he is simply the

The roaring or 'belling' of the rutting red stag echoes in the autumn night. The normally timid animal now shows much more aggression both towards rivals of his own species and other animals daring to invade his territory.

Facing page: The red stag's antlers are shed every year in early spring and start to regrow immediately. Towards the end of July the downy enveloping cover of skin, known as velvet, flakes off and the antlers remain bare until they once more drop. The stag's neck thickens and reaches maximum dimensions in the rutting period. Development of antlers is more rapid in mature adults than in young, immature stags.

strongest stag in the neighbourhood but this dominant position does not imply any responsibility on his part for group surveillance and protection. It is possible that such a well endowed male may make an attempt to intimidate weaker members of the clan but as a general rule he will allow the younger animals to precede him, following at a prudent distance.

The significant differences in the behaviour patterns of the male and female herds are particularly manifest when any kind of danger threatens. The highest ranking hind will immediately show acute signs of disquiet and give vent to her suspicions by uttering a series of crisp warning barks which automatically put her companions on the alert. Retreat and flight, if necessary, are efficiently organised as the whole group moves off in single file. The dominant hind leads the way and the one next in order in the hierarchy brings up the rear of the procession.

The stags, in contrast, give no audible warning signals to one another when an enemy is scented. Although they too sometimes adopt the single file pattern of retreat it tends to be more disordered. Position in the column seems to depend on how quickly each stag responds to the emergency and where he happens to be when the group starts to move.

In summer red deer make for the cooler hilly districts and usually come down onto the sheltered plains for the winter. Many of the stags, however, elect to brave the cold in more exposed areas. In any event the two types of herd still stay well apart. Territorial frontiers do not seem to be designed to resist the intrusion of an animal from another herd but simply to mark the limits beyond which no member of the community, particularly in the case of the hinds, ventures to stray.

Rut and reproduction

Autumn is the rutting season and the only time of year when this rigid clan system undergoes change. The stags leave their summer quarters and settle down alongside the hinds. All signs of male solidarity disappear as each reproductive takes up residence on his rutting ground, his objective being to round up as many of the hinds as he can. The stag announces his ownership of territory by rending the silence of the September night with his clamorous roaring.

Although they appear to accept the domination of the lord of the harem, the hinds retain their matriarchal social structure and form no kind of permanent relationship with him. Should it happen that this newly constituted group is forced to take flight, the dominant male gallops away at top speed, paying not the least heed to the safety of those partners that were recently claiming his full attention. The latter are quite capable of fending for themselves and move off in an orderly file.

Not all the stags show a desire to procreate at the same time. First to show interest are the oldest individuals and they mark out their territorial boundaries some time before the younger stags. The maximum period that a reproductive male will spend in the company of the hinds is a month but he is not likely to remain master of his own harem and territory for more than about one week. Continual efforts to keep his hinds together in a group, repeated confrontations with neighbours trying to invade his domain and failure to eat properly while under stress of sexual excitement all combine to weaken him. Soon the embattled stag is unable to ward off his rivals and is forced to yield his place to one of them. Once expelled from his territory the stag makes no immediate effort to dislodge an even feebler individual. He makes for a quiet, remote spot where he can spend several days feeding in order to restore his strength.

These refuges are very often situated fairly close to those belonging to similarly evicted stags but the animals show complete indifference to one another which rules out the possibility of any fighting. In due course the exhausted, starving animal has recovered his weight and energy and is now ready to return to the fray and challenge a momentarily weakened rival for possession of the latter's harem.

With the termination of the breeding season the reproductives part company with the hinds and reform into the small groups which will remain undisturbed until the following autumn. Now the brockets or yearlings come into rut but make no attempt at this point to stake a claim to territory or venture in search of a partner. At best a few couple with stray hinds.

Antlers and tines

In winter the red stag leaves his sheltered hilltop site in the evening and wanders down into the valley in search of food. One morning in March, replete from his nightly meal, he may happen to catch his head against a low-hanging branch across the narrow path that leads back to the plain. There is a dry cracking sound

The antlers of the Irish elk of the peat bogs measured more than 9 feet across and their enormous size and weight may have led indirectly to the extinction of the species in the Pleistocene epoch. The heavy antlers would have prevented the deer escaping the attacks of predators and primitive hunters.

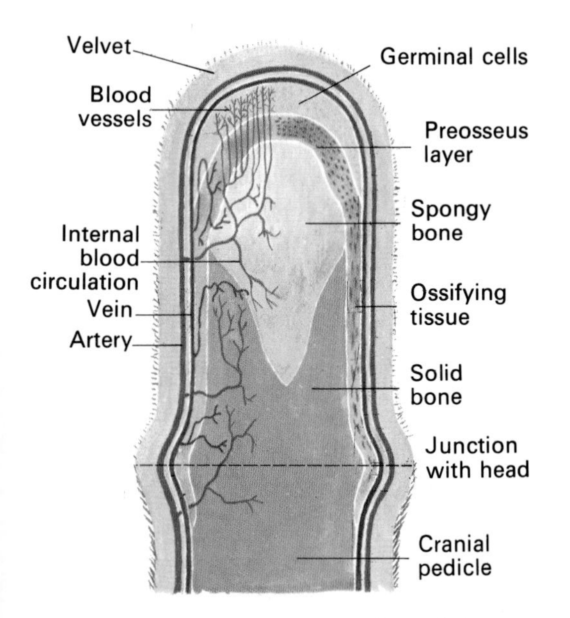

Longitudinal cross-section of a red deer's antler.

Facing page (above) : Stags often meet in combat to dispute possession of hinds and territory but the antlers seldom inflict serious injury. (*Below*) A belling stag with hinds and calves.

as his antlers snap through at the base. The startled deer bounds off into the mist.

This is of course no accidental occurrence but a perfectly normal natural phenomenon whereby the males (of this and of most other deer species) shed their antlers once a year and later renew them. Yet although it is a simple enough process it poses interesting questions for zoologists and not all of them have so far been satisfactorily answered. They know how, but not why it happens and they are not in agreement as to the relative advantages of this annual shedding and regrowth.

Although deer have always been common enough in the Holarctic region not much attention was paid until comparatively recently to the precise nature and function of the antlers adorning the heads of the males. From the hunter's point of view they might be no more than splendid trophies to hang on the wall but obviously nature never intended them to be purely decorative adjuncts. It has been left to naturalists to throw more light on this intriguing problem.

Structurally the antlers of a deer are formed of solid bone, without a marrow cavity, sprouting directly from pedicles on the frontal bone of the head. The developing bosses or knobs are rounded (later coming to a point) and their growth is conditioned by the periosteum, a double layer of fibrous tissue which is richly furnished with blood vessels. Another important element in antler growth and regrowth is the food eaten by the deer, containing large proportions of calcium and phosphorus.

Yet these factors are not alone responsible for the annual replacement of cast-off antlers. The vital additional substance is the sexual hormone secreted by the testes, known as testosterone, circulating in the bloodstream. It has been shown that if a stag is castrated its antlers will not develop in the normal manner, remaining as they were at the moment of the operation. But as soon as the castrated animal is injected with the hormone which it is no longer able to supply of its own accord, the antlers begin to grow once more.

So much for the composition of the antlers and the factors that determine their development. But what exactly is their purpose? The hinds or does of the majority of deer species do not possess them; nor do the males of two smaller members of the Cervidae, the Chinese water deer and the musk deer (in compensation they have well developed upper canines). Yet antlers are not the sole prerogative of the stag or the buck for when we consider the European reindeer and the American caribou we find that both males and females carry antlers. So quite clearly these are not simply secondary sexual characteristics analogous to a man's beard or a lion's mane. Furthermore, it is believed that the Irish elk, the primitive giant deer of the peat bogs, became extinct because of the enormous size of its antlers. These measured more than 9 feet from tip to tip and weighed something like 150 lb, possibly handicapping the animal's movements to such an extent that it fell victim to all manner of predators. This theory might seem to lend support, therefore, to the view that a deer's antlers are really aberrant growths, a mistake made by nature.

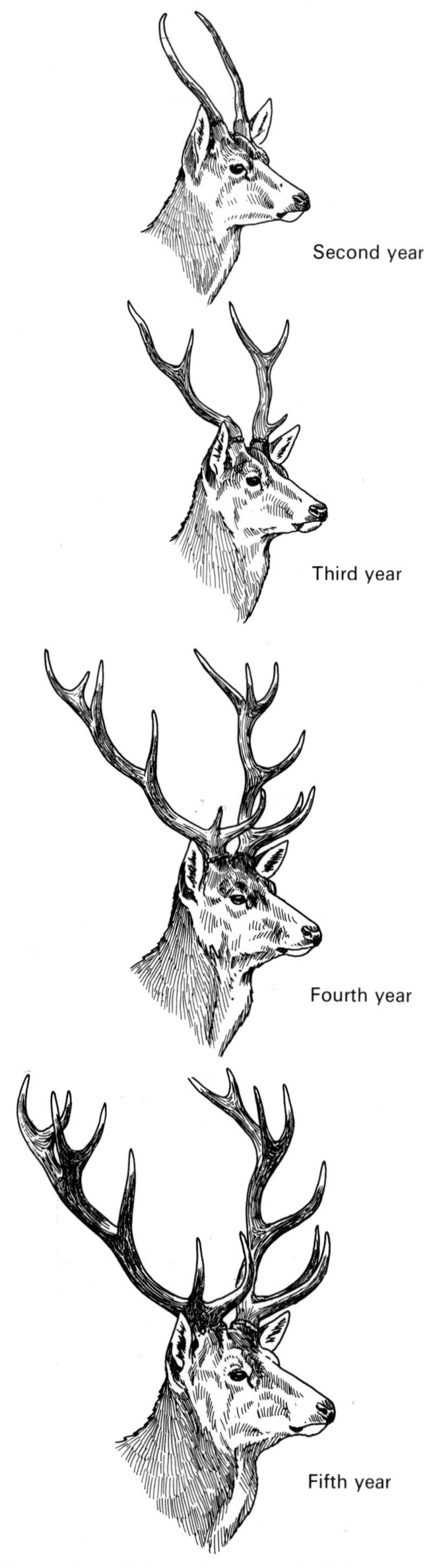

The antlers of the red stag reach their maximum size and development in the seventh or eighth year. They continue to drop and regrow annually but later begin to degenerate.

Yet might not this apparent handicap prove to be a means of assuring the survival of a species? When deer fight among themselves the main objective appears to be to dominate and conquer a rival without necessarily causing him severe injury. The heads thrust forward, the antlers interlock, but the aim is rather to force a retreat, to dispossess the opponent of his territory and harem, not to kill him. In the last resort this rival is a member of the same community who will be needed for procreation the following season and whose untimely death could prove to be a serious loss to the herd. This is why the combat must be relatively harmless and ritualised, the antlers serving principally as weapons of dissuasion. Their main purpose is thus to repel rivals in the course of intraspecific fighting and this could explain why the females do not have need of them. The antlers of the males are at their strongest during the rutting season, when most of these combats take place, but they still persist well into the winter so that they can if need be perform a defensive role against predators.

There is a fundamental difference between the antlers of deer and the horns carried by many other ruminants. Deer antlers are deciduous (dropped completely) and renewed each year, unlike the horns of cattle, sheep, goats and antelopes. Most people are aware of this distinction but there is an amusing story of one dealer in wild animals who was ignorant of the fact, to his cost. Packing a consignment of African antelopes destined for the United States, he carefully cut off their horns so that the animals would fit into the cages provided, confident that they would regrow on arrival. Unfortunately antelope horns are permanent and once broken off are lost for ever.

In the case of deer, the shedding of the antlers causes a small haemorrhage and shortly afterwards a layer of cicatricial tissue appears in the form of a scar. Unlike the skin growing over a wound, however, this is covered with hair. The new antlers start to sprout almost immediately but look different from the discarded pair, being enveloped in a downy sheath of velvet, this skin layer being well furnished with blood capillaries carrying the necessary growth elements and criss-crossed by a network of sensitive nerve fibres. When the antlers are fully developed the dried velvet falls off in strips and to make the process easier the deer rubs its head repeatedly against tree trunks.

As the stag gets older the number of branches or tines on the antlers gradually increases though not, as some people think, at the rate of one a year. The animal's age can be roughly determined by the state of development of the antlers, although this can vary according to its diet and other factors. The original knobs are replaced by simple spikes in the second year and in the third year these are replaced by stronger growths known as beams, with a brow tine near the base and a tine near the top called the trez. While the beam continues to grow, rough formations known as pearls develop near the base. In its fourth year the animal is of the 'fourth head' and now carries an additional tine above the first, known as the bez tine. At five or six years of age the animal becomes a 'five-pointer', each beam now being furnished with at least five tines, the three at the top forming the

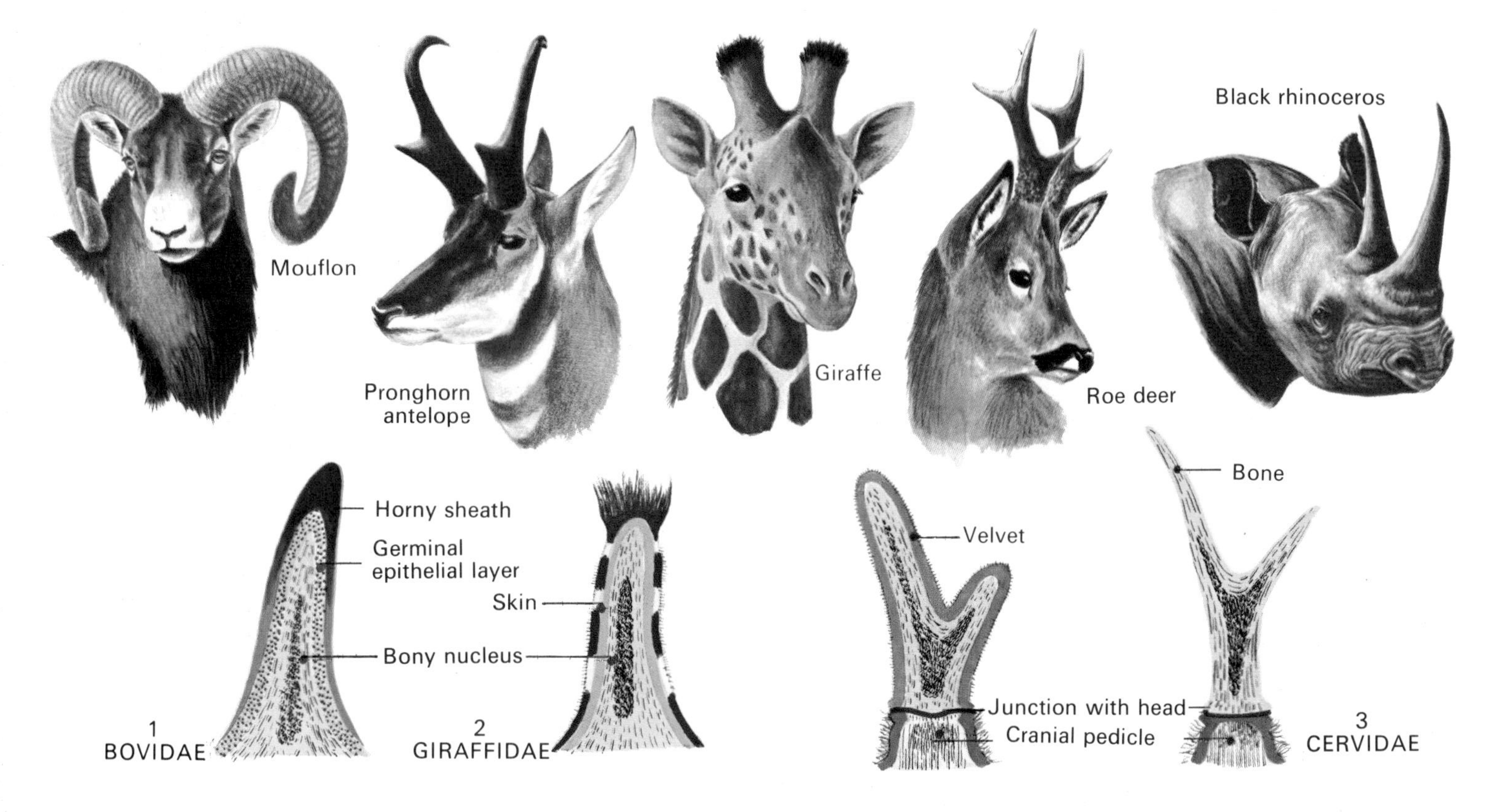

Only five families of mammals–Bovidae, Antilocapridae, Giraffidae, Cervidae and Rhinocerotidae–possess horns. Strictly speaking, the term applies only to cattle whose horns consist of a bony core enveloped in a horny sheath (1). In the pronghorn antelope it is only this outer sheath which is shed and renewed annually. The bony antlers of the deer family break off completely each year and grow again, being covered with soft velvet which dries and flakes off when antler growth is complete (3). The Giraffidae occupy a half-way position between cattle and deer; the okapi sheds its small layer of enveloping skin but the bosses of the giraffe are permanently covered with hair (2). The so-called horns of the rhinoceros are purely epidermal.

royal antler. Two years later the stag is a 'large five-pointer'. But when the deer is from ten to twelve years old the glandular mechanism stops functioning properly. The antlers regrow annually but in a random fashion and are in fact degenerating.

The dappled fawns

The fawns of the red deer, usually called calves, are born in late spring or, especially in northern regions, in early summer. The mature hinds account for about 60 per cent of the births, for the younger females, fertile in their third year when conditions are favourable, normally give birth only every other year.

Shortly before she is due the mother-to-be wanders away from the herd, accompanied by previous offspring, and settles in a sheltered spot well out of range of her companions. She has already parted from her mate who by this time has gone off with other males.

For the first few days the newborn calf, with its dappled coat that blends so perfectly with the surrounding vegetation, makes no attempt to move, while the mother keeps attentive watch nearby. Whenever she prepares to suckle her offspring she takes infinite precautions, approaching tentatively and making several détours so as not to disclose the whereabouts of the baby. It is during these critical days when the calf is incapable of following its mother that up to 50 per cent of newborn deer succumb to predators such as lynxes, golden eagles, wildcats and, above all, foxes. According to Fraser-Darling the three or four pairs of raptors in the region he was studying killed at least half a dozen newly born calves each year. But if a hind temporarily separated from her calf scents danger she will come to the rescue immediately, attacking the enemy with her hooves, sometimes driving it off, sometimes dying in the attempt.

Calves that survive these perilous days are soon able to venture out on their own and are often seen gambolling with elder siblings or with youngsters of the same age. They seem to have certain favourite games, one of which consists of a fawn clambering up to the top of a hill or cliff, rearing up on the hind legs and inviting others to try dislodging it. Alternatively, the frisky youngsters will chase and challenge one another, kicking out with their fore-feet but never actually landing a blow. At other times an entire group may dash madly around a hillock until one of them takes possession of the summit, at which point they all come to an immediate halt.

By autumn the dappled coat has turned to plain brown and the antlers of the males are sprouting. Later they are known by various names. A second year calf is called a brocket, in its fourth year it is a staggard and in the fifth year a hart. At six years old it is a fully grown stag.

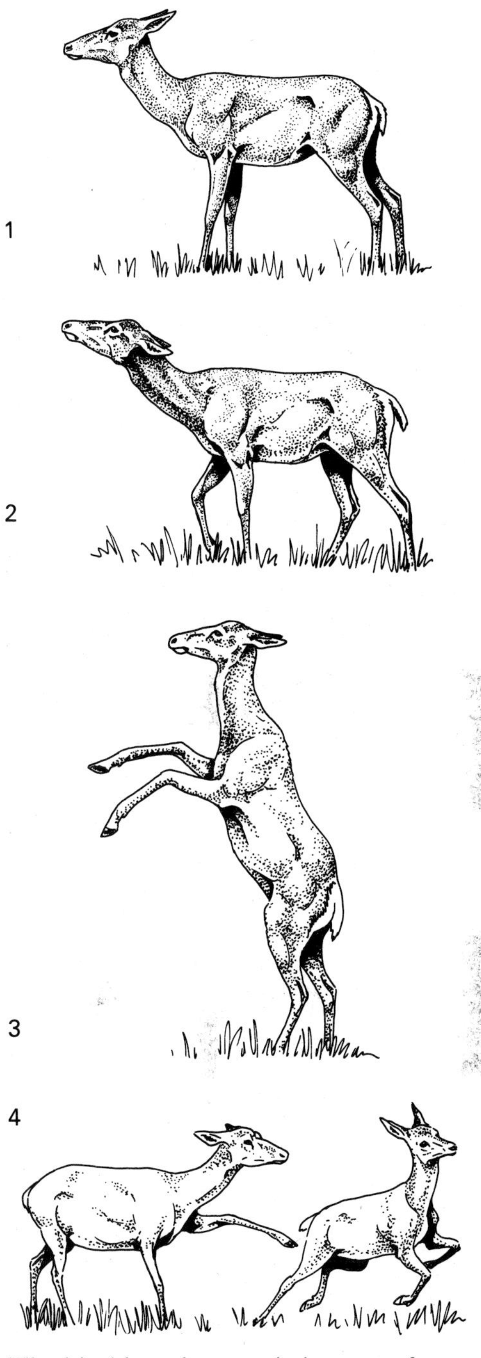

The hind has characteristic ways of expressing her moods–warning (1), threatening (2) or imminently aggressive (3). If annoyed by her offspring she may graze its rump lightly with the tip of her hoof (4).

Facing page: Shortly before she is due to give birth the hind searches for a quiet, secluded spot in the undergrowth. She will not leave her baby until it is capable of standing and following her back to the herd.

Enemies large and small

At one time the principal enemy of the red deer was the wolf, a pack of these predators being capable of killing a splendid five-pointer. Wolves are nowadays rare in Europe and as a result

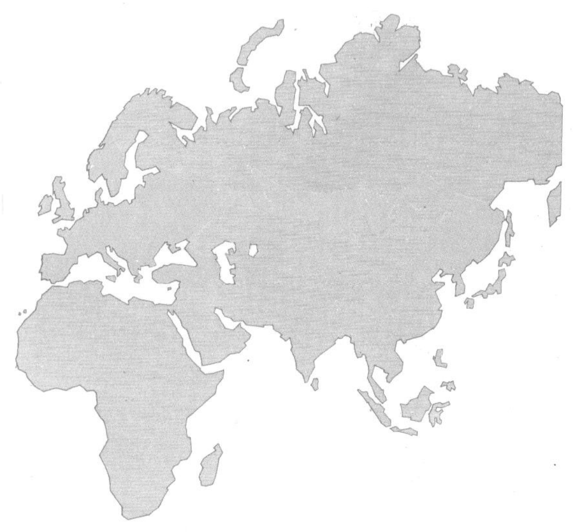

Geographical distribution of the roe deer.

the red deer population has proliferated, so that hunting controls have been necessary to keep the numbers down and thus prevent the animals destroying their habitats. This is more sensible than the ill-fated operation in the Kaibab forest of Arizona when local hunters killed all the predatory pumas in the neighbourhood, only to find that the surviving deer gradually deteriorated in quality and eventually perished for lack of food. We now know that the activities of natural predators are necessary for the well-being and survival of these species. Wolves, for example, hunt in such a way that they can obtain the largest amount of protein with the least expenditure of energy. They kill only young, sick, incapacitated or aged animals, eliminating individuals that would otherwise pass on hereditary defects to their progeny, as well as impotent old stags that nevertheless attempt to monopolise the hinds. Man, on the other hand, uses his gun to pick out the most magnificent and virile members of the herd and thus his hunting activities could be said to run counter to the workings of natural selection.

The red deer's most formidable predators include the wolf, the fox, the puma, the bear and the golden eagle but there are a number of much smaller, insidious enemies which, although not directly fatal, weaken the deer to such an extent that they become easy prey for their traditional foes. Many kinds of insects parasitise deer, some of which announce their presence by buzzing around their victim. This throws the deer into a panic. It shakes its head violently as if seized by a fit of madness and gallops off to higher ground where its torturer is unable to follow. Other insects, however, are more inconspicuous, laying their eggs in the pelage of the unsuspecting host without causing any discomfiture.

The life cycle of the latter parasites is interesting because they spend much of it actually within the body of their host. The deer botfly of the genus *Cephenomyia*, for example, flies around the nasal openings of the ruminant and with the aid of strong buccal hooks lays 500–600 eggs inside the nostrils. In February, when about 20 millimetres long, the larvae work their way down into the deer's pharynx and continue growing there until they are double that length. In March the deer, irritated by these foreign bodies, begins sneezing violently and ejects the parasites. But once the latter are in the open they develop a protective envelope and within three or four weeks emerge as adults, capable of procreating and initiating a new life cycle. The warble fly (*Hypoderma diana*) lays her eggs in the deer's pelage, causing no immediate irritation. The larvae, however, burrow deep into the deer's skin and muscle tissues. In January they are infesting the animal's dorsal region and three months later they dig their way out and fall to the ground, there to complete their development.

ROE DEER
(*Capreolus capreolus*)

Class: Mammalia
Order: Artiodactyla
Family: Cervidae
Length of head and body: up to 54 inches (135 cm)
Length of tail: 1–$1\frac{3}{4}$ inches (2–4 cm)
Height to shoulder: 30 inches (75 cm)
Weight: 35–65 lb (15–30 kg)
Diet: herbivorous
Gestation: 165 days
Number of young: 1, 2 or 3

Adults
Small, graceful deer with hind legs longer than front pair and rump usually higher than shoulders; tail hardly visible. Only the buck has antlers, about 9 inches long, seldom with more than three tines. They drop in November and have regrown by February, velvet starting to be shed in March. Winter pelage is greyish with a white mark on the throat and an even more conspicuous one on the rump. In summer the coat is reddish-brown and the head greyish (especially in the buck). Size is variable as is the date of antler regrowth. Siberian roe deer may weigh over 100 lb.

Young
Newborn fawn has reddish coat with dappled white marks in three lines along the back.

The roe deer

The roe deer (*Capreolus capreolus*), much smaller than the red deer and far less gregarious, has adapted more successfully to human presence. This small ruminant with its short tail, long ears and slender legs, has a greyish coat in winter which turns red in summer. Its average weight is about 65 lb although in Siberia it may exceed 100 lb. Extremely abundant in Europe and Asia, its favourite habitat is dense valley and mountainside undergrowth. Because of its cryptic coloration, however, the roe deer is not easy to locate. In one forest in Denmark, for example, the resident roe deer population was estimated at around 70 but later proved to be nearer the 200 mark.

There have been no natural obstacles to the steady extension of the roe deer's range which today includes domestic pastures and farmland. The woodland thickets bordering these areas have sufficient shade for shelter and the cereal crops constitute particularly choice food. Because of this and also as a result of the reduction in numbers of natural predators in recent years, the roe deer population has risen significantly. It has been established that the species has experienced dramatic fluctuations in numbers over the centuries. Few fossils or cave drawings of the animal have been discovered from the Paleolithic age so that it seems unlikely that the roe deer was the object of intensive hunting at that early period. But in Neolithic times there was evidently a marked extension of its distribution range. Later there must have been a population recession, followed quite recently by a recovery. The variations are apparently due to successive modifications of the deer's forest habitats.

The roe deer is ideally built for a woodland life. Its comparatively small size enables it to hide effectively in thickets. The short antlers–rarely exceeding 9 inches in length–do not interfere with progress through the undergrowth. Large ears provide it with exceptionally keen hearing, vitally important in a habitat where visibility is poor. The hind legs are longer than the front legs, making it easier to climb or jump over fallen trunks and other obstacles in its path.

In some European countries there are more roe deer than ever before. In Sweden, for example, within the last 150 years the numbers have risen from about 100 to over 100,000! This staggering increase has posed serious problems for the forestry authorities because of the damage done to young trees.

Contrary to what was once believed, roe deer exhibit strong territorial instincts during the rutting season. Each buck selects its piece of rutting ground and jealously defends its frontiers against all comers. Whenever a couple of neighbours meet on the common borderline they give vent to their fury and the ensuing battles are in some cases so fierce that one of the rivals may be killed. The bucks are so aggressive at this season that they may work off excess energy on nearby trees, destroying them completely. With a large roe deer population the damage may thus be quite extensive.

Faced with such a situation the governments most affected decided to put an end to this destructive activity by launching an extermination campaign. This was easier said than done, partly because the animals themselves were too adept in concealment and partly because the authorities came under fire from protest groups of conservationists who were not prepared to stand by and watch the wholesale destruction of a species.

Like other members of the Cervidae roe bucks shed and renew their antlers annually. Although smaller and with fewer tines than those of the red deer, they grow in the same way, complete with velvet.

Stages in the courtship ritual of the roe deer.

Facing page: Although the red deer is the most noble of forest animals the roe deer is much more graceful. The buck controls the activities of his little family group and the doe normally gives birth to one or two kids.

The British naturalist Richard Prior hit upon a compromise solution designed to protect both the animals and their woodland environment. As a member of the Forestry Commission this gifted student of wildlife was given responsibility for supervising the local roe deer population and quickly learned to become familiar with the habits of every animal in his district. In the course of his work he discovered that the bucks were extremely conscious of their territorial rights in the breeding season. He also demonstrated that the area occupied by a buck was directly related to its size. The larger the deer the more extensive his territory and vice-versa. It seemed to him that the damage done to young trees was much more dependent on the actual number of bucks in the vicinity than on the weight and strength of the dominant males. Thus the most practical method of maintaining the deer population at an acceptable level was, in his opinion, to eliminate the smaller animals, not to massacre an entire herd indiscriminately. His recommendations were acted upon and the results were completely successful, proving that it was possible to achieve the double objective of preserving the forests and safeguarding their animal inhabitants.

The secret rites of the roe deer

Roe deer live in family groups consisting of a buck, a doe and fawns, usually called kids, of less than a year old. This little society, unlike that of the red deer, is patriarchal in structure. In winter two or three families sometimes come together but such associations do not last long. At the beginning of spring the groups separate, each to its own territory.

In May the does that were impregnated during the previous summer isolate themselves in the thickets, far out of sight of prying eyes. There they give birth to one, two or (on rare occasions) three kids. The number of offspring depends on the age of the mother. A doe giving birth for the first time usually has only one kid; later she may have two or exceptionally (and only when she is a good deal older) three. The newborn kids weigh 2 lb and have reddish hair, dappled with white.

In the case of a multiple birth the members of the family do not all remain together but put some distance between one another, ensuring that if predators are lurking they cannot discover and kill more than one kid at a time. Shortly after they are born, after the mother has carefully licked them clean, the babies make shaky efforts to get to their feet and within two or three hours have succeeded. After about a week they have trebled their birth weight and are already browsing on leaves. A week later they have again doubled their weight and at three weeks they are ruminating. The doe stays close to her offspring for at least the first month, but as soon as they are capable of following her around she abandons her refuge in the undergrowth and explores much farther afield.

During the day roe deer normally remain inside the wood but will venture out either at dawn or at dusk to feed in the clearings or on the forest fringes. In summer, prior to the harvest, they enjoy resting in the cornfields where they can be free of disturb-

Stages in the ritual combat of rival roe deer.

ance. They feed on mushrooms, leaves and the shoots of poplars, willows, birches, oaks, service trees and other deciduous species, as well as on pine needles in spring. When they are adult they require a daily quantity of food equivalent to about three-quarters of their own weight.

The antlers of the buck drop and begin to sprout anew in late winter (kids only acquire their first rudimentary pair in the second year) and by February they are fully developed. They are three-tined, weigh a little more than a pound and are still enveloped in velvet. This covering dries and is shed in April (or earlier in some southern regions) revealing the hard rough bone of the antlers underneath.

There is now a marked alteration in behaviour. In winter, although not exactly sociable, the deer were prepared to tolerate the presence of others of their kind. They would rummage together in the snow for tufts of grass and buried leaves or they might lie in small groups on the southern slopes of a hill, warming themselves in the sun. Others would huddle together near a stream or under the bare, wind-swept branches of an oak, a poplar or an ash.

Now the days gradually grow longer and the sun gives out much more warmth. The fury of the north wind has abated and there is a scent of mountain freshness in the breeze which ruffles the beech branches on which buds are beginning to appear. The ground is soft underfoot and the undergrowth bright with new season's greenery and flowering plants. As the forest comes alive once more and spring gives way to early summer the territorial instincts of the bucks are again aroused and soon they are making ready for the breeding season.

The oldest bucks, guided by experience, stake their claims first, choosing their territory in a well sheltered spot as close as possible to water and with plenty of food to hand. By the time the younger mature males appear on the scene there are few choice sites that have not already been reserved by means of scent posts. Rebuffed right and left, they eventually settle for a smaller, poorly located patch or wander off into the depths of the forest.

As the rutting period approaches – usually in July and August but often extending into September – the bucks become markedly more aggressive. Only a short while previously the small deer would start at the slightest noise and immediately take to their heels but now they are fierce in the defence of their territory and will attack intruders. Cases have even been reported of roe deer charging and injuring humans who have unwittingly invaded their domain. The bucks trot after the does, the latter pursuing a complicated zig-zagging course through the woods and across plains and fields. Muzzles thrust forward, the bucks follow close behind and the tracks of the animals are deeply imprinted in the ground as they describe circles and figures-of-eight round trees, tufts of grass and rocks. These paths are commonly known as roe rings.

After the sexual act there is a delay in the implantation of the blastocyst so that although nine months may elapse between mating and birth, gestation lasts only about five months.

Facing page : There are representatives of the order Artiodactyla in all the typical biomes of the Holarctic region, as shown in this chart. Some have adapted to more than one type of habitat.

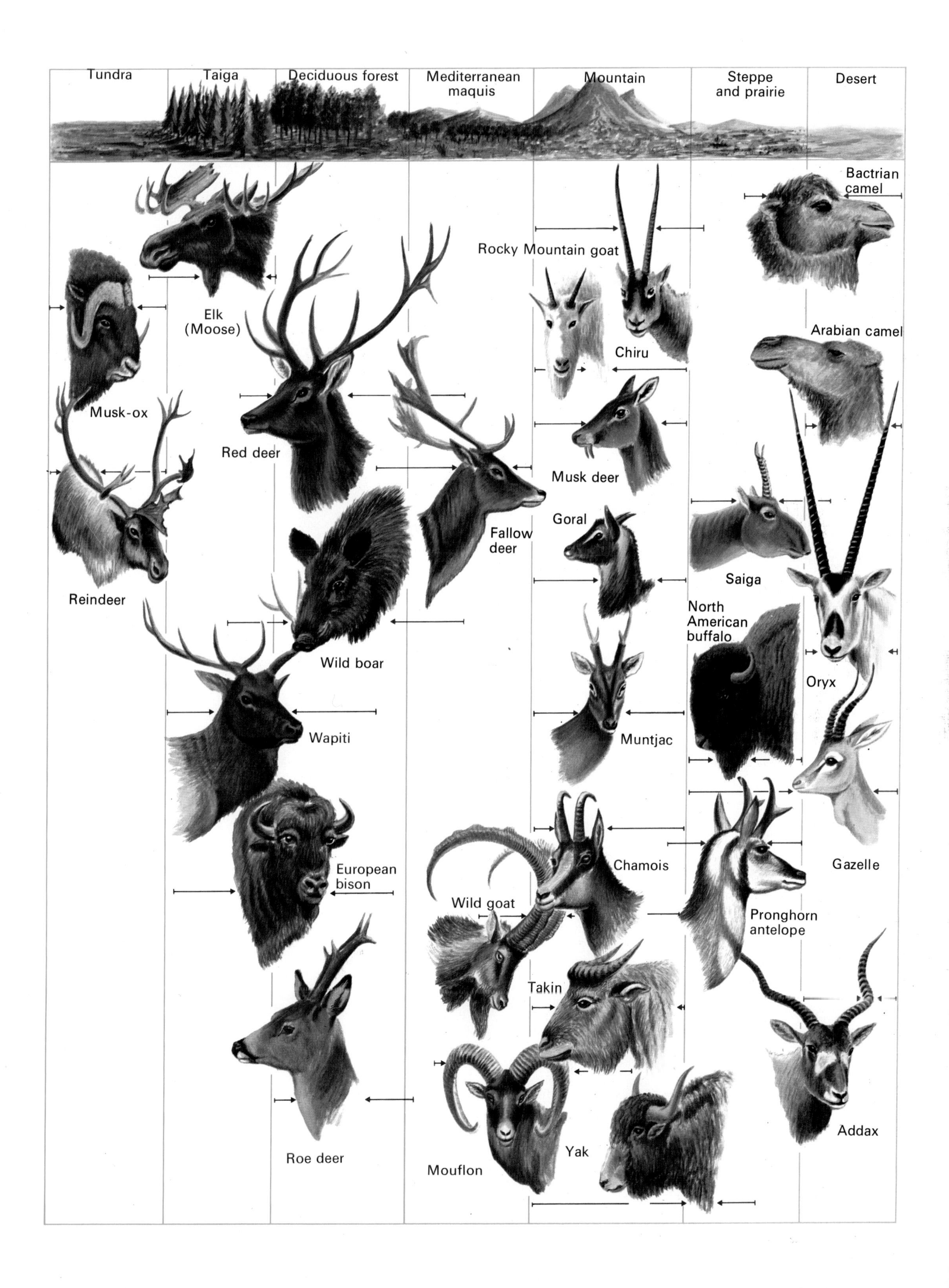

Tundra
Taiga
Deciduous forest
Mediterranean maquis
Mountain
Steppe and prairie
Desert
Elk (Moose)
Musk-ox
Reindeer
Red deer
Fallow deer
Wild boar
Wapiti
European bison
Roe deer
Rocky Mountain goat
Chiru
Musk deer
Goral
Muntjac
Chamois
Wild goat
Takin
Mouflon
Yak
Bactrian camel
Arabian camel
Saiga
North American buffalo
Oryx
Gazelle
Pronghorn antelope
Addax

FAMILY: Cervidae

The deer family–Cervidae–belong to the order Artiodactyla (even-toed mammals) and are represented by many species living in North and South America, North-west Africa, Europe and Asia.

The various kinds of deer inhabit a wide range of biomes that include swamp, tundra, taiga, deciduous forest, Mediterranean maquis and open steppe.

The majority of these animals tend to band together in compact herds but some of them, such as the elk, lead a solitary existence except during the breeding season.

The most characteristic physical feature of the family is the presence of bony growths sprouting from the head. These antlers vary in size and shape and are found in all genera with the exception of *Moschus* and *Hydropotes,* represented respectively by the musk deer (*Moschus moschiferus*) and the Chinese water deer (*Hydropotes inermis*). As compensation for lack of antlers these deer possess long, curving upper canines or tusks that protrude downwards from the mouth. The muntjac or barking deer (*Muntiacus muntjac*) has both antlers and tusks.

Antlers are exclusively male attributes, with the exception of the two species of genus *Rangifer*–the reindeer (*Rangifer tarandus*) and the caribou (*Rangifer arcticus*)–in which they are carried by male and female alike. It is not unknown for the occasional female of other deer species to sprout antlers but these are invariably small and deformed.

The antlers usually appear in the first or second year when they are unbranched. As the animal gets older they increase in size and complexity of structure, with numerous tines.

The deer sheds its antlers regularly each year and then grows a new pair. In temperate regions they usually drop off between January and April but in warmer countries the phenomenon may occur at any time of year.

Shortly after they drop off the antlers begin to grow again, being covered with a soft layer of skin known as velvet. The dried velvet falls away in strips, being finally shed when the antlers are fully grown.

With deer living in warm climates there may be several rutting times a year, whereas those inhabiting temperate or cold regions have only one rut a year, either in autumn or winter. As a general rule one or two fawns are born but the number may sometimes be three or even four.

The newly born fawns of many species are dappled with white but this later gives way to a uniformly coloured pelage. The dappled coat is retained as an adult, however, by the fallow deer (*Dama dama*)–though only in the summer–and by the axis or Indian spotted deer (*Axis axis*).

There is considerable diversity in size and weight among the Cervidae. The musk deer weighs around 20 lb whereas the largest member of the family, the moose (*Alces americana*), stands 6 feet at the shoulder and may weigh more than 1,750 lb. There are also notable variations within a single species, depending on the surroundings. In general the females are smaller than the males, with a more graceful shape, principally because the neck is not so thick and muscular.

All deer are herbivores, feeding on grass, leaves and sometimes lichen and bark. The stomach is four-chambered.

The dental formula is:

$$I: \frac{0}{3} \quad C: \frac{0\text{–}1}{1} \quad PM: \frac{3}{3} \quad M: \frac{3}{3}$$

Many species have been extended well beyond their original range through being introduced by man to distant countries such as Hawaii, New Guinea, Australia and New Zealand, even South Georgia in the Antarctic–regions where no artiodactyls had previously existed.

Musk deer
(Moschus moschiferus)

Chinese water deer
(Hydropotes inermis)

The only two deer species without antlers are the musk deer and the Chinese water deer, both of which have distinctive protruding fangs.

Facing page: The white-tailed or Virginian deer (*Odocoileus virginianus*) and the sika (*Cervus nippon*) are just two of the many members of the far-flung deer family, whose range covers the Holarctic, Oriental and Neotropical regions.

CHAPTER 10

The crafty hunters of the forest

The struggle for survival in the deciduous forests of the Holarctic region may not be so obvious or spectacular as it is on the African savannah, principally because of the absence of really large predators such as the lion, the leopard and the cheetah. But the grim game of life and death goes on just the same. Even though the action is less conspicuous and the actors themselves of smaller dimensions the rules do not change.

Among the more prominent of forest carnivores, the wolf, the lynx and the puma have all but vanished from their traditional habitats in Europe. The most powerful bird of prey, the golden eagle, has also been swept from the skies. Their place in the food chain has lately been usurped by the common or red fox (*Vulpes vulpes*). Since man has helped to remove its former enemies and chief competitors, the fox has come into its own. Today it roams both forest and farmland, though not with impunity.

The fox—fact and fable

Renard the fox has acquired a colourful reputation in story and folk-lore and is popularly regarded as being sly, cunning, deceitful, malicious and destructive. As so often happens when imagination is allowed to run riot, truth is grossly distorted. Yet the fables are quite correct in emphasising that the fox is an uncommonly intelligent animal. Laboratory experiments have shown that its reactions are in many ways superior to those of a dog. Its hunting strategy, for example, is a compound of absolute familiarity with the lie of the land, rapid decision making and an excellent memory. Yet for lack of first-hand information certain aspects of the fox's behaviour are still misunderstood. When poaching was much more prevalent than it is today, the fox

Facing page: The red fox, a beautiful member of the canine family, is a swift and silent hunter of forest and woodland, active at all seasons, including winter.

Food preferences of the red fox.

would frequently filch a victim from a snare and deposit a pile of excrement at the scene of the crime. Poachers arriving too late to claim their prize seriously argued that this was the carnivore's way of expressing its contempt for them; but this was before naturalists had discovered anything about animal territory and scent posts. Such gaps in knowledge have now been filled but other mysteries remain.

The red fox is a member of the family Canidae and looks much like a bushy-tailed dog. The fur is usually a rich reddish-brown, the outsides of the legs and the backs of the large ears being black and the underparts white. The really distinctive feature is the long tail or brush which normally has a white tip although this may sometimes be black.

Examination of the stomach contents of foxes has proved that they enjoy a very varied diet. One German survey revealed that about 82 per cent of all ingested food consisted of animal protein (of which more than half were small rodents) and approximately 18 per cent of different vegetable substances, principally fruit. Apart from grapes and berries the omnivorous foxes absorb any vegetable substances contained in the digestive tracts of their prey. The list of victims is headed by rodents, followed by young and adult hares and rabbits. Next in order are birds, reptiles, insects and other invertebrates, and fishes. These forms of prey are supplemented by a high proportion of carrion.

It is astonishing, therefore, that an animal which preys mainly on pests such as rodents and which also feeds on putrefying carcases should have the reputation of being uncommonly

COMMON OR RED FOX
(Vulpes vulpes)

Class: Mammalia
Order: Carnivora
Family: Canidae
Length of head and body: up to 40 inches (100cm)
Length of tail: 12¾–20 inches (32–50 cm)
Height to shoulder: 14–16 inches (35–40 cm)
Weight: 9–33 lb (4–15 kg)
Diet: omnivorous
Gestation: 50–63 days
Number of young: generally 2–6, sometimes up to 8
Longevity: 10–14 years

Adults
Short muzzle, large pointed ears; brush usually white-tipped. Overall colour reddish, underparts white.

Young
Newborn cub has soft, thick, dark hair.

destructive, notably to livestock. Extermination campaigns would be organised on the flimsy pretext that dead lambs had been discovered in foxes' dens or along paths they regularly frequented. On the few occasions that scientists had the opportunity to examine such remains they were able to prove conclusively that most of the victims had been dead before the predator arrived to drag them away to the lair. But although declarations of war against foxes often turn out to do more damage than they are designed to prevent, there is no getting round the fact that some individuals do occasionally cause havoc in the farmyard. As far as poultry are concerned, the surest defence against nocturnal marauders is for hen-houses to be well constructed and care taken that the foxes cannot burrow under the wire.

The European fox population has extended its range well beyond its original habitats and the animal is often seen on open ground or even in gardens on the outskirts of towns. So fearless of man has it become that earths (dens) have actually been found in factory precincts, military encampments and sports stadia.

Foxes frequently roam around the fringes of woods to capture mice and other small rodents. As soon as they sight their prey they stop and remain quite still, like a pointing dog.

Territory of the dog-fox

Except during the breeding season the male or dog-fox prefers to avoid the company of others of the species, communicating only from a distance. Thus he may announce his presence either by barking or by setting up scent posts, utilising scent-secreting glands. The glands situated between the sole pads are apparently

The fox's den is known as an earth and it is here that the cubs are born in spring. Since it is often no more than the former refuge of a badger or a rabbit, there is no nest, at most only hair from the vixen's underside. The cubs' thick fur, however, protects them from the elements. As the pictures on the facing page show, the fur is initially dark and later becomes red.

used for tracing itineraries while those located at the root of the tail give out a particularly sharp odour in the rutting season and, together with excrement and urine, serve mainly to mark territorial frontiers. This is what is happening when a fox is seen rubbing its hindquarters against the trunk of a tree.

The female (vixen) also makes use of scent posts but in her case it would appear to be more a matter of social status than defence of territory. Several vixens may in fact share the same hunting ground, although at different times of the day or night. In the course of such expeditions they may emit barks in order to make their presence known and to avoid meeting other vixens.

The yelps of the dog-fox during December and January are certainly designed to let potential rivals know that he has taken possession of territory. Every night will then be spent patrolling his domain, always following an identical route. Each fox has his own distinctive vocal register and if there are a number of animals in the neighbourhood each can be identified by the tone and rhythm of his bark. The characteristic 'foxy' smell is very much in evidence at this season.

It is at the end of the summer that the cubs born the previous spring attain their independence and venture out to establish their own earths. The chances are that all the choicest sites will already have been occupied by their elders and there may be fights which are extremely noisy but not really violent. The younger dog-foxes are invariably banished to more sparsely populated districts. This mass exodus of inexperienced young animals is not in the best interests of the species for many of them are caught in traps (which, unlike their elders, they have not learned to avoid) or are killed by hunters.

As a general rule one or more vixens may take up residence inside the territory of each dog-fox. Although they have sometimes been said to be polygamous, foxes are monogamous, and the dog-fox plays a large part in the rearing of the cubs, contrary to the popular belief that he has nothing to do with them. When a vixen dies–and with hunters on the trail this is a frequent occurrence–the dog-fox will continue to feed them, provided they have been weaned. In countries such as England, where hunting is a popular sport, the death rate among vixens may be very high, many of them not surviving beyond the age of two years.

The fox's bushy tail or brush serves as both balancer and rudder so that the animal can brake or turn quickly in mid-course with surprising ease and agility.

Parents and cubs

Most of the mature vixens come on heat in January although they may sometimes be seen mating a little earlier or later. But no matter when it occurs, she is sexually responsive only for a short period of 24–36 hours. It is during the breeding season that the animals are most frequently seen and the dog-foxes often venture out in broad daylight.

Very little is known about the reproductive behaviour of the species and much of the existing information is still the subject of controversy among naturalists. The sexual act is preceded by a kind of dance in the course of which both animals bound around each other, sometimes in silence, sometimes emitting growls. The pre-mating play is very like that seen in the cubs. After this

ritual performance the foxes mate. As with most members of the dog family, copulation lasts several minutes, the male becoming 'knotted' or 'tied', as in the mating of the domestic dog.

By the end of January these activities are usually concluded. The dog-foxes are less active, the barking dies down and the musky scent is less noticeable.

Some time before the foxes mate the vixen chooses her earth, generally a hole or burrow belonging to another animal but which she may occasionally dig herself. An abandoned rabbit hole is ideal, as is a badger's sett. It was once believed that badgers found foxes' effluvia intolerable but since vixens have been observed living harmoniously with them as neighbours at this time of year the theory appears unfounded. It is probable, however, that in some instances the badger removes itself as soon as the fox cubs are born. Once she has taken possession of her earth the vixen remains there throughout the winter, giving birth at the end of March or the beginning of April.

It has been said that the dog-foxes prefer to sleep in the open but this is highly unlikely.

The dark fur of the newborn cubs is soft and thick, forming a necessary protective cover since the vixen provides no lining to the earth other than the hair she pulls from her underside, which has the advantage of uncovering her teats, so helping the cubs to suckle. For the first month they are fed exclusively on milk and the vixen remains close at hand, rarely leaving the earth, even to go hunting. A little later she will settle down outside the den, watching over the cubs as they explore the immediate neighbourhood. By now they are old enough to progress to solid food brought to them either by the dog-fox or by the mother.

By May the vixen only returns to visit her cubs at sunrise, in the middle of the afternoon and at sunset. At dusk she approaches the earth with extreme wariness, following a circuitous and complicated route and silently making off at the faintest sound. If she scents the odour of a human along the path she begins barking loudly, the signal being received by the cubs who immediately retreat to the depths of their earth. The daytime visit is attended by even greater precautions and it is very difficult to follow her movements.

The patient observer can reap no finer reward than a glimpse of the cubs innocently gambolling in the fresh air, chasing one another about, as if pretending to hunt. In their case, as with most other predators, such games are in some ways preparations for adult experience. Nevertheless, one should not be tempted to exaggerate their significance. Some authors, for example, insist that the vixen deliberately buries part of her prey so that the youngsters can discover it. This seems unlikely and if the cubs do dig up buried food it would appear to be accidental.

By June the litter will be living outside the earth, sheltering in the undergrowth and feeding mainly on invertebrates. The vixen now visits the family infrequently. The end of August sees the cubs almost independent, making only brief contact with their mother. It is now that the dog foxes start to stake out their territory or roam far afield in search of an unoccupied site. Ritual combat may occur between September and November

COMMON MUSTELINES

Class: Mammalia
Order: Carnivora
Family: Mustelidae

PINE MARTEN
(Martes martes)

Length of head and body: $16\frac{3}{4}$–$21\frac{1}{4}$ inches (42–53 cm)
Length of tail: $8\frac{3}{4}$–$10\frac{3}{4}$ inches (22–27 cm)
Height to shoulder: 6 inches (15 cm)
Weight: 2–$5\frac{1}{2}$ lb (1–2.5 kg)
Diet: omnivorous, mainly flesh
Gestation: 8–9 months
Number of young: 2–5

Long body, large head, pointed muzzle and ears. Soles of feet covered with hair. Long tufted tail. Colour mainly dark brown, yellowish-orange below neck; border of ears creamy-white. In summer the coat is lighter.

BEECH MARTEN
(Martes foina)

Length of head and body: 16–$20\frac{1}{2}$ inches (40–52 cm)
Length of tail: $8\frac{3}{4}$–11 inches (22–28 cm)
Height to shoulder: $4\frac{3}{4}$ inches (12 cm)
Weight: 6 lb (2.6 kg)
Diet: omnivorous, mainly flesh
Gestation: 8–9 months
Number of young: 2–4

Similar in appearance to pine marten, with smaller, more slender ears. Colour dark brown; neck and border of ears white.

POLECAT
(Mustela putorius)

Length of head and body: $12\frac{1}{2}$–18 inches (31–45 cm)
Length of tail: $4\frac{3}{4}$–$7\frac{1}{2}$ inches (12–19 cm)
Weight: 1–$2\frac{1}{2}$ lb (500–1200 g)
Diet: flesh
Gestation: 6 weeks
Number of young: 3–6

Long, stout body, flattened head and rounded muzzle; short, bushy tail. Overall colour dark chestnut. The face looks like a black mask with white borders to the ears, two white stripes above the eyes and white lips. The female is more slender than the male. The young are pale grey and do not have the black mask or white markings.

Facing page : The red fox has greatly extended its original forest range and often ventures into the open, quenching its thirst in pond, lake or stream.

but there is seldom any bloodshed. The females born the previous year tend to come on heat a little later than the older vixens and are already capable of bearing an initial litter towards the end of their first year.

Opinion is divided about the precise role played by the dog-fox in the rearing of the cubs. Some authors claim that he is totally unconcerned with their welfare while others affirm that he is extremely attentive during these early days and that he brings the family food at regular intervals. This latter view is supported by those who have actually seen dog-foxes–presumably the fathers of the litters concerned–visiting the earth from time to time and remaining outside for a lengthy period, as if mounting guard.

Hainard, who has made a thorough study of the behaviour of the red fox in Switzerland, also describes the movements of a tailless dog-fox (and thus easily recognisable) which continually brought back prey in his mouth, depositing it at the entrance to an earth occupied by a vixen and several cubs.

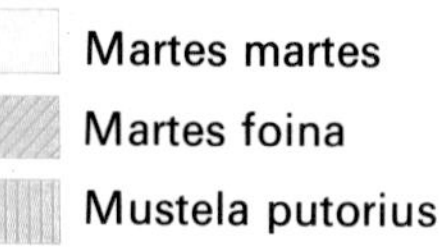

Geographical distribution of the pine marten (*Martes martes*), beech marten (*Martes foina*) and polecat (*Mustela putorius*).

Weasels to wolverines

The red fox is an animal of the woodlands although its predatory activities extend to open country and urban areas, a steadily growing range of distribution for which man, by eliminating the astute carnivore's larger rivals, is partially responsible. Other hunters of the forest–even more silent and secretive–have benefited in the same way from the disappearance of traditional predators. These are the Mustelidae–weasels, stoats, badgers,

martens, minks, polecats, otters and wolverines–sharp-witted animals, most of them with streamlined bodies and short legs which, though seldom seen, play an extremely important role in the ecology of the deciduous forests of Europe. Because of their strong carnivorous tendencies they not only regulate the smaller herbivore population but also assist man by destroying rodents.

These animals vary considerably in size, ranging from the tiny weasel, less than a foot long, which can squeeze through the narrowest gaps, to the voracious wolverine or glutton, whose total length is over 4 feet and which will attack any animal, however large. The Mustelidae hunt on the forest floor, below ground, high up in trees and under water, each animal with its own specialised hunting technique and favourite prey.

Although most of these fierce carnivores confine themselves to dense woodland, some of them will explore built-up areas. Beech martens, for example, have adapted very successfully to life in towns and city suburbs; and where these animals have been allowed to roam at will there has been a marked decline in the local rodent population. Unfortunately it is far more common for weasels and martens to be tracked down and killed on the pretext that they destroy certain game animals and are thus detrimental to man's interests. They may indeed play havoc with rabbits in some regions (though since the onset of the myxomatosis epidemic this has happened much less frequently) but their marked preference for rodents gives them a utilitarian value that greatly outweighs their misdeeds. Farmers who urge wholesale campaigns of extermination, mindless of the elementary laws of ecology, are merely damaging their own interests.

The agile carnivores of the family Mustelidae are well represented in the deciduous forests of the Holarctic region. Here, from left to right, are the ermine (stoat in winter livery), a pair of badgers, the weasel and the beech marten. All these animals have a wide food range, with rodents high on the list, except in the case of the badger.

The aggressive tendencies of polecats appear at a very early age and have been closely studied by T. B. Poole who has shown how early games take on a serious significance. A polecat will pursue its selected protagonist (1), try to drag it along the ground (2) and to bite its neck (3). Adults engaged in intraspecific combat, however, will often launch attacks from one side (4) and adopt menacing attitudes, baring their teeth and emitting sharp cries (5). In the latter case the objective is rather to expel the rival than to injure him. The adult polecat (*below*) hunts all kinds of small mammals, birds, frogs and lizards.

Typical mustelids of the deciduous forest include the pine marten (A), beech marten (B), polecat (C), otter (D), mink (E), ermine (F), badger (G) and weasel (H). Food and habitats are sufficiently varied for them not to enter into direct competition with one another.

Play with a purpose

The polecat (*Mustela putorius*) may be regarded as the prototype of this carnivorous family for it exhibits all the most characteristic anatomical features and habits of the group. With its long supple body and very short legs it can wriggle its way into narrow clefts, burrows and tunnels, giving the occupants no chance of escape. Although a formidable forest hunter, most active where vegetation is fairly sparse, since it is a poor climber, it will easily adapt to a variety of habitats and situations and is frequently found on the banks of streams and rivers, being expert at swimming and diving.

The aggressive tendencies of the polecat are evident from an early age, as has been attested by the naturalist T. B. Poole. The games which young polecats play with one another when they are only four weeks old already have a serious undertone and reach their climax within the next month. The antics begin when one animal leaps onto the back of a companion and tries to bite its neck. The victim rolls on the ground and defends itself by sinking its sharp little teeth into the face and neck of the adversary, the latter doing its best to dodge the blows. If one of the animals gains a stranglehold on the other desperate measures may be needed in order to break free. At other times the aggressor will relent and the two polecats will then switch roles.

The young polecat seems to have an irresistible urge to find an outlet for its aggressive feelings. If unable to find a willing

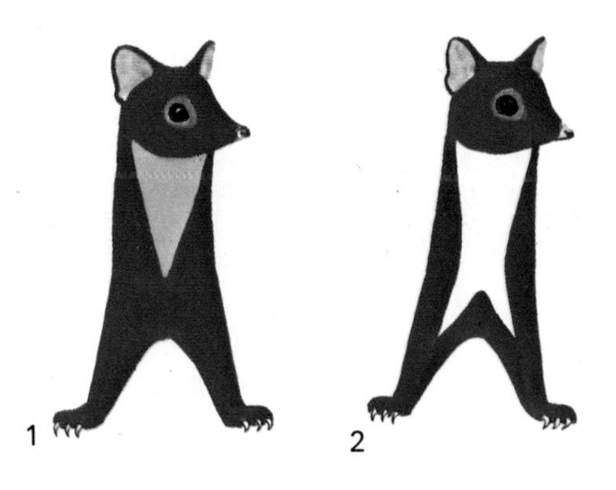

The pine marten (1) and beech marten (2) are distinguished by their throat markings.

The red squirrel is found in many European forests but is now fairly rare in Britain.

partner for its rough game it may leap around another member of the same litter, teasing, provoking and biting. When the reluctant object of these taunts has had as much as it can endure it begins to hiss and show its teeth, thus putting an end to the irritating antics of the mischief-maker.

The uncommonly early emergence of such behaviour (which may sometimes be evinced in newborn animals even before they open their eyes at about two weeks) indicates that it is purely instinctive. Despite claims to the contrary, there is no possibility of the babies having learned to act in this way, however valid the apprenticeship theory may be for other species. In its playfulness the young polecat has much in common with the majority of other predators. It is to be hoped that more detailed observations, either in the wild or in captivity, will throw more light on this fascinating phenomenon.

Red squirrel
(Sciurus vulgaris)

Grey squirrel
(Sciurus carolinensis)

Woodland acrobats

The best known forest animal and a familiar resident of parks and gardens too is the squirrel. Attractive to look at, entertainingly acrobatic, readily tamed and sociable, it is understandably popular although too few people recognise it for what it is–a rodent with very destructive habits. Sentiment, however, will have its due. The bright little eyes, perky ears, bushy tail, frisky movements and human-style method of eating food from the hands, are certain guarantees of sympathy and affection.

Squirrels are highly sensitive to temperature changes. If it soars above 25°C (77°F) they cease their activities; in winter they do not hibernate and although they may spend more time in the nest their tracks can always be seen in the snow.

The squirrel population seems to fluctuate enormously in five- or six-year cycles and also appears occasionally to be migratory. In the Soviet Union the animals have been seen undertaking journeys of more than 200 miles, in the course of which they pass through towns and–even more surprisingly–cross rivers and sounds. They are at all times astonishingly agile, scuttling head-first down tree trunks and making 15-foot leaps from branch to branch. Hainard once saw a young squirrel lying temporarily stunned after jumping from a height of about 60 feet to the ground. It soon recovered and scampered up another tree. The toes (four on the forefeet, five on the longer hind feet) and sharp claws give firm support and the tail is used as parachute, balancer and rudder. There are in fact the beginnings of a flying membrane on the flanks.

The two common European species are the red squirrel (*Sciurus vulgaris*)–now rare in Britain–and the grey squirrel (*Sciurus carolinensis*). They feed mainly on acorns, beech mast, nuts, fir cone kernels and the like, cracking husks and shells with their teeth and claws, and laying up reserve stocks for the winter. In addition they eat flowers and berries, insects, eggs, fledglings and occasionally carrion. Their unpopularity stems not only from the range of their food but also the damage caused to tree bark.

Squirrels enjoy rolling about in wet moss. It is said, but unlikely to be confirmed, that if surprised at such activities they will hurl nuts at the intruder.

During the courtship ritual the male positions himself alongside the female, then moves round her in a circle, balancing with his tail and erecting the hairs of his back. When she is receptive she chases him, letting out cries similar to those used by the young to summon their mother. During the breeding season the female occupies a nest on her own, driving the male away should he come near. In the north the animals mate during the spring, in the south between January and April and from May to August. The babies are born naked and blind, opening their eyes after 32 days. The male departs as soon as they are born and within six weeks the youngsters are able to leave the nest.

The squirrel hunters

Acrobatic though they are, the squirrels are hardly a match for their two principal forest predators–the beech marten (*Martes foina*) and the pine marten (*Martes martes*). The former is not found in Britain and their respective names give a clue to their normal habitats. Similar in appearance and habit, the pine marten confines itself to the depths of the forest and shuns human company, unlike the beech marten.

The slender marten is just as agile as its squirrel prey, leaping audaciously from branch to branch, even though they bend precariously under its weight, and making use of its tail as a rudder in mid-air.

Once it has located its victim the ensuing chase will be fast and furious. The squirrel skims like a tightrope-walker along the thinnest branches in the hope that in this way it will throw off its pursuer. The marten's answer to this manoeuvre is to take an

The squirrel scampers up and down tree trunks with extraordinary agility and can make long leaps from branch to branch. Although a popular little animal its claws do much damage to bark.

Geographical distribution of the red squirrel (*Sciurus vulgaris*) and grey squirrel (*Sciurus carolinensis*).

SQUIRRELS OF THE HOLARCTIC REGION

Class: Mammalia
Order: Rodentia
Family: Sciuridae
Diet: mainly vegetarian, but some flesh
Gestation: 46 days
Number of young: 1–7 usually 3–4

RED SQUIRREL
(Sciurus vulgaris)

Length of head and body: 7½–11¼ inches (19–28 cm)
Length of tail: 5½–9½ inches (14–24 cm)
Weight: ½–1 lb (230-480 g)

Long tail and tufted ears. Pelage red, apart from belly which is white, and white fringe to tail in winter in British subspecies.

GREY SQUIRREL
(Sciurus carolinensis)

Length of head and body: 9¼–12 inches (23–30 cm)
Length of tail: 7½–10 inches (19–25 cm)
Weight: ¾–1¾ lb (370–750 g)

Larger than red squirrel; no tuft on ears. Upper parts grey, underparts white.

Facing page: The pine marten can match the squirrel for agility and speed and is thus the latter's most formidable enemy. It is much sought by hunters and trappers for its fur.

outflanking path to cut off the squirrel's escape and thus force it down to a lower level in the tree. As the trapped squirrel makes a desperate leap to safety the marten lunges upwards to intercept the victim in mid-air, seizing the smaller animal with its sharp claws, dragging it to the ground and rolling over and over until it finds an opportunity to sink its canines in the squirrel's throat.

The marten can tackle larger animals equally effectively by implanting its long teeth in the neck of the prey and breaking the spinal column.

All kinds of small and moderate sized mammals as well as birds, fishes, amphibians and invertebrates are hunted by the marten according to availability and it is very partial also to eggs, honey and fruit.

The marten will rarely attack domestic animals and it has only an insignificant effect on species normally hunted as game. Unfortunately, however, its rich fur is in great commercial demand so that there is still an adequate excuse for shooting it in many regions where it is still found in numbers.

The marten is a territorial animal which roams a domain of up to 10 miles radius. Given its speed the animal can cover this area from end to end in a single night. It normally takes refuge in the dreys of squirrels, in the nests of crows and birds of prey, in hollow tree trunks, in rock clefts or in abandoned lairs and burrows.

Most births occur in April and May, the litter of the pine marten consisting of 2–5 young, that of the beech marten of 2–4 babies. Covered with whitish fur when born, their development is slow and they suckle until they are six weeks old, abandoning their refuge at about this time.

The European wildcat

Living in the depths of the forest but with a very reduced range in comparison with some centuries ago, the European wildcat (*Felis sylvestris*) is an extremely ferocious animal. But because it keeps well out of man's way hardly anything is known about the habits of this, the most independent representative of the cat family.

Once found in many parts of Britain, it is nowadays confined to the Scottish Highlands, but its principal strongholds are rocky woodland regions in southern and eastern Europe.

This cat normally hunts by night although during the autumn it sometimes emerges from its hiding places in the daytime. The hunting grounds are usually extensive and are fiercely defended by the male.

Most authors take it for granted that the animal is strictly monogamous in view of the fact that pairs are frequently seen together. Some individuals apparently lead a solitary, roving existence. The rut may occur either at the beginning of March or towards the end of May and early part of June. Four or five kittens are born either in May or August.

There is considerable controversy about the relationship of the European wildcat and the ordinary domestic cat (*Felis catus*). It is true that the former looks like a large tabby but experts

nowadays incline to the view that the domestic cat is descended from the African wildcat or bush cat (*Felis lybica*) which is comparatively docile and has sometimes been tamed. But since there is not all that much difference in appearance between the African and European wildcats other zoologists believe that both animals may belong to the same species (there have in fact been cases reported of successful cross-breeding). Nevertheless, when it comes to behaviour they are far removed from each other and all attempts to domesticate *Felis sylvestris* have completely failed.

The ways of the raccoon

The raccoon (*Procyon lotor*) of the North American forests is a delightful animal with a greyish-brown body, a bushy ringed tail and a dark mask-like band across forehead and eyes. A group of raccoons feeding by the edge of a stream puts one in mind of tiny dwarfs, an impression reinforced by the fact that they use their hands for conveying food to the mouth.

These animals are classified among the Procyonidae and are thus related to the pandas and kinkajous which utilise their forepaws in a similar fashion. The sense of touch is in fact extremely well developed and they make use of their hands for all exploratory activities.

The raccoon is an intelligent animal and laboratory experiments have established that its grasp of a situation is far superior to that of the cat or dog, in fact closely approaching that of certain monkeys. Pushing open a door with its hands, lifting up a lid to see what is inside a box or untying a knot are just a few typical actions which present no problems for the inquisitive raccoon. George Heinold claims to have watched one amusing little animal unscrewing the covers of several pots of honey and feasting on the contents.

One of the habits which has greatly fascinated zoologists is that of dropping morsels of food in water and puddling them, as if washing them. If one presents a lump of sugar, for example, to a raccoon in captivity it immediately plunges it in its drinking trough and shows itself much perplexed as the titbit gradually disappears. Experts have tendered several explanations for such strange behaviour. For some time it was widely assumed that it was essential for the animal to moisten all food substances so that they would be soft enough to swallow, the reason given being that the salivary glands were atrophied. Subsequent autopsies, however, established that these organs are normally developed. The suggestion was also made that the raccoon's fingers were more sensitive when slightly wet. Other authors pointed out, however, that food 'washing' was not the normal custom in the wild and seemed to be confined to individuals in captivity. They based this judgment on observations of raccoons in the wild which would eat their food dry when there was no water in the vicinity and would soak it only when feeding close to a river or lake. But this still left unexplained the fact that baby raccoons reared in zoos immediately moisten their food as soon as it is presented to them by their keeper.

Thanks to their varied diet and flexible behaviour patterns,

African wildcat
(*Felis lybica*)

Domestic cat
(*Felis catus*)

European wildcat
(*Felis sylvestris*)

raccoons have adapted successfully to different habitats, including those environments that have been artificially created by man to serve his own needs. Although originally forest and woodland inhabitants they now find cultivated farmland equally acceptable.

Hunting, fishing and refuse collecting

Raccoons hunt by night. In spring their diet is almost entirely carnivorous and they hunt rabbits, squirrels, birds and young beavers. They also use their climbing skill to steal eggs and fledglings from nests built high in trees. In summer, however, 70–80 per cent of their food consists of vegetable matter, the balance comprising crabs, frogs and other amphibians. On sea coasts they gather clams and oysters, scooping them from the shells with astonishing dexterity, and in some regions they feed on turtles, often consuming up to 50 per cent of their eggs.

Autumn is the season when raccoons devote almost all their time to finding food for they have to build up suitable reserves for the winter. Acorns, nuts and other dry fruits, as well as insects are consumed in large quantities.

Rivers and streams are rich sources of food and even when the water is muddy and turbulent this does not prevent raccoons from fishing. Since they cannot actually see their prey they cleverly immerse themselves in the water, waiting immobile until a fish brushes against them. Then they lash out with a paw,

Facing page : The origin of the domesticated cat is a matter of controversy. Some experts claim that it is descended from the European wildcat, pictured here, a highly aggressive animal whose habitat is confined to inaccessible wooded mountain regions. Most trace a line of descent from the more sociable, more easily tamed African wildcat.

An inhabitant of the North American forests, the raccoon is a highly adaptable animal with a varied food range. It makes good use of its forepaws and flexible fingers for climbing, stealing fledglings from the nest, catching fish and conveying all food to its mouth. It frequently dips food in water as if washing it, but the precise significance of this habit is not known.

Geographical distribution of the raccoon.

RACCOON
(Procyon lotor)

Class: Mammalia
Order: Carnivora
Family: Procyonidae
Length of head and body: 28–33½ inches (70–84 cm)
Length of tail: 8–12 inches (20–30 cm)
Height to shoulder: 8–12 inches (20–30 cm)
Weight: 11–15 and up to 35 lb (5–7 and up to 16 kg)
Diet: omnivorous
Gestation: 60–70 days
Number of young: 2–7
Longevity: 10–12 years

Stout body, small neck, pointed muzzle, triangular, slightly rounded ears. Colour mainly greyish-brown; typical black mask round eyes. Tail greyish-yellow with 5–7 black rings.

Facing page: As well as being an efficient hunter of ground animals, the raccoon is an expert at fishing. It does not even need to see its prey, simply stationing itself at the water's edge and waiting for a fish to brush against its paw.

seldom missing their objective.

In winter, when they are relatively inactive, raccoons eat practically everything they can find, including carrion. The carcase of a deer, for example, is a real prize. If they happen to be near a town or farm they will even overturn dustbins to sort through the contents. Many people deliberately leave scraps of food for them in the garden. Leonard Le Rue cites the case of one American family which awaited the arrival in their kitchen, punctually each evening, of a small group of raccoons that proved to be as tame as any household pets.

Maternal responsibilities

Raccoons mate in January or February, the males being polygamous and fighting fiercely among themselves for possession of the females. According to Le Rue, however, it is the latter who make the final choice of partner and – contrary to what is generally assumed – they do not necessarily grant their favours to the males that have proved their superiority in combat.

After mating the female falls into a semi-lethargic sleep which lasts until spring while her partner continues to be normally active. She then prepares a nest, usually in the trunk of an old tree, and after a gestation of 60–70 days gives birth in April or May to a litter of from two to seven young. The latter weigh a couple of ounces and already possess the distinctive facial mask of the adults. They open their eyes at about ten days but do not leave the nest until they are eight to ten weeks old. The father plays no part whatsoever in rearing them and is in fact repulsed should he make a tentative effort to do so. There are several confirmed cases of males who have eaten their offspring in the temporary absence of the mother. She is normally extremely attentive and prudent, changing the nest should there be any threat of disturbance and carrying her babies one by one in her mouth to the new place of refuge.

The young are suckled for about fourteen weeks. By the time they are two months old they weigh about 2 lb and are beginning to climb. But although they are soon capable of getting about on their own they stay close to their mother for approximately a year, spending their time playing, climbing tall trees or simply sitting on the ground imitating the adults. The mother supervises all activities and soon steps in should the games become too rough or one of the youngsters be injured.

The principal enemies of the raccoon are nocturnal predators such as the bobcat, puma and wolf. But although it is only of moderate size, the raccoon is not a compliant victim and will defend itself with extraordinary bravery. The compact body, sturdy neck, powerful muscles and thick fur are all physical assets which enable it to fight back with grim determination.

Unfortunately no amount of courage is of much avail against man's traps and guns. The early pioneers of North America (including the celebrated Davy Crockett) used raccoon skins and tails as personal adornments; and the animal's attractive fur is still prized by hunters and trappers, in response to a steady demand from the garment trade.

Typical representatives of each of the five subfamilies of the Mustelidae.

FAMILY: Mustelidae

The representatives of the family Mustelidae – one of the most important and complex subdivisions of the order Carnivora – are mammals of small or moderate size. Among characteristic anatomical features are an elongated body, short legs that usually terminate in five toes with sharp claws and, in many species, highly developed anal glands with strong-smelling secretions. Some of these animals are digitigrade, others semi-plantigrade or plantigrade.

Those members of the family that lead an aquatic life have toes joined by a membrane, which forms a kind of swim paddle.

The typical dentition formula is:

$$I: \frac{3}{3}\ C: \frac{1}{1}\ PM: \frac{2\text{-}4}{2\text{-}4}\ M: \frac{1}{1\text{-}2}$$

The majority of mustelids are predators attacking a variety of land animals but some of them, such as the otter, feed principally on fishes, whereas others, such as the sea otter, consume large quantities of molluscs. The badger is an omnivore.

The head of a typical mustelid is slender, the muzzle comparatively short and the ears small and rounded. The skull is broad and flattened. The length of tail varies according to species.

There is still much to be discovered about the reproductive habits of these animals. The period of gestation, for example, is very variable, ranging from about six weeks in certain species to eight or nine months, even more than a year, in others. It has also been established that, as in the case of one ungulate, the roe deer, there is a delayed implantation. That is, the fertilised ovum undergoes division to form a blastocyst and this takes some time to attach itself to the wall of the uterus.

Different animals belonging to the same species may come into rut at different times so that it is hard to determine exactly when the mating season occurs. Studies with ferrets have shown, for example, that, in common with other animals, the degree of sexual activity in the female is directly related to the respective durations of daylight and darkness. Nor is there any great consistency in the number of young per litter.

Some representatives of the family are nocturnal hunters, others are diurnal. The weasels are essentially ground dwellers, the badgers are burrowers, the martens live in the trees. Others lead an aquatic existence either in streams, rivers, lakes and marshes, or on sea coasts (otters), and in the sea itself (sea otters).

On the basis of their anatomical differences and diverse behaviour patterns, zoologists have broken down the Mustelidae into five subfamilies– Mustelinae, Mellivorinae, Melinae, Mephitinae and Lutrinae.

The Mustelinae have an enormous distribution range. The animals in this group are on the small side, most of them with digitigrade feet and markedly curved claws. There are ten genera. The stoat, polecat, ferret, weasel and mink belong to genus *Mustela*; the martens to genus *Martes*. The glutton or wolverine, a plantigrade resembling a small bear, is the only representative of genus *Gulo*. The African zorillas, which project their anal secretions for some distance, belong to genus *Ictonyx*. The remaining six genera are *Vormela* (marbled polecat), *Tayra* (tayra), *Grison* and *Lyncodon* (grisons), *Poecilictis* and *Poecilogale* (muishonds).

The Mellivorinae contains a single species, *Mellivora capensis*, the ratel or honey badger, with a range that embraces almost the whole of Africa south of the Sahara and much of southern Asia, from Arabia to India. It is a heavy, sturdy animal, not unlike the European badger but with a different dental structure. Whereas the common badger has four premolars in each half of both jaws (though one tooth in the upper jaw tends to fall out early in life), the ratel has only three.

The subfamily Melinae include the various badgers of Europe, Asia and North America–burrowing animals with long thick hair, a relatively heavy body, pointed muzzle and short tail. The common badger (*Meles meles*) is an inhabitant of Europe and Asia. The sand or hog badger (*Arctonyx*), the teledu or Malayan stink badger (*Mydaus*), the Palawan stink badger (*Suillotaxus*) and the Bornean ferret badger (*Melogale*) are confined to the Far East. The American badger (*Taxidea*) is similar to the common badger but its head is not so definitively striped.

The skunks of the subfamily Mephitae are mammals of North and Central America, with long, thick black-and-white fur. The anal glands secrete a nauseous-smelling substance which can be sprayed for several yards to deter an enemy. There are three genera–*Mephitis* (striped and hooded skunks), *Spilogale* (spotted skunk) and *Conepatus* (hog-nosed skunk).

The representatives of the subfamily Lutrinae have adapted to life in and near water and are characterised by a long body, a strong rudder-like tail and webbed feet. The members of genus *Lutra* (common and Canadian otters) live in rivers, lakes and swamps in Europe, Asia, America and North Africa. The clawless otter (*Aonyx*) is an inhabitant of the marshlands of East and West Africa. The sea otter (*Enhydra*), as its name suggests, is an offshore animal, found specifically along the coasts of the North Pacific. Other otters belong to the genera *Pteronura* (Brazilian giant otter), *Amblonyx* (Indian small-clawed otter) and *Paraonyx* (African small-clawed otter).

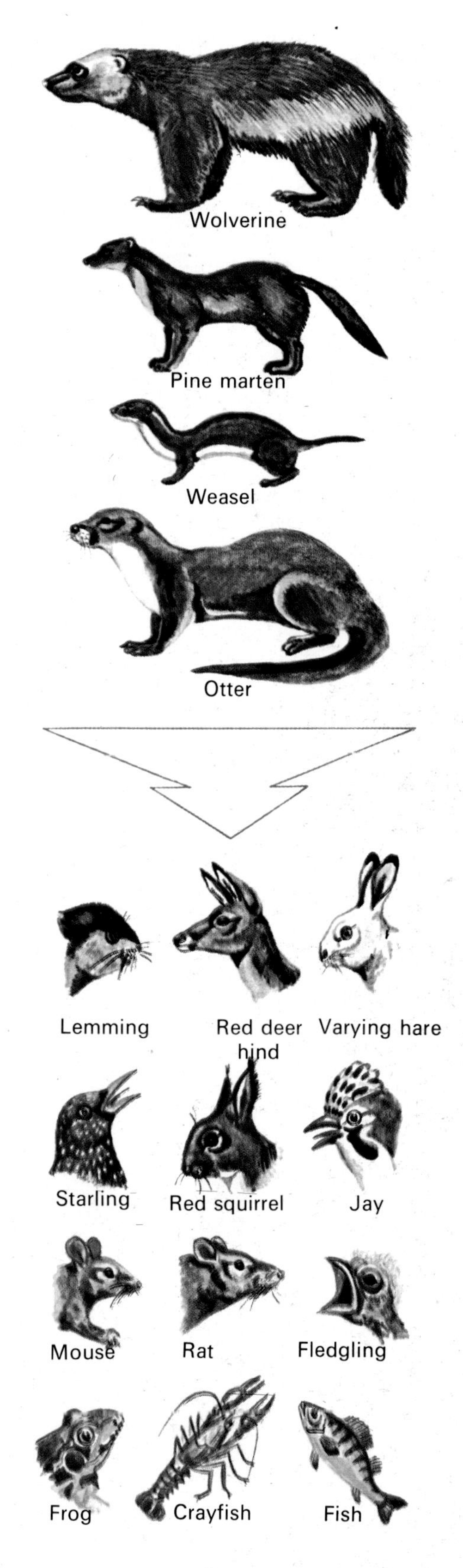

As shown in this chart, the food range of the various species of mustelids is extremely varied, including mammals, birds, fishes and amphibians.

CHAPTER 11

Marauders of the woodland glades

Birds of prey are less numerous in the deciduous forests of Europe, Asia and North America than in more open, spacious biomes, for the simple reason that dense tree cover is detrimental to their normal activities of aerial pursuit and combat. Closely packed trees, tangled branches and thick foliage are of no assistance to the eagle scanning the ground for prey and are positively dangerous to the peregrine falcon as it nose-dives earthwards at high speed. Yet these forests harbour a rich variety of game that, in spite of the risks involved, smaller raptors cannot afford to ignore; and some of these birds have adopted specialised hunting techniques so effective that no furred or feathered inhabitant of wood or forest can feel entirely safe from attack.

Two of the most highly specialised woodland hunters operating by day are the goshawk (*Accipiter gentilis*) and the sparrowhawk (*Accipiter nisus*). Although these birds belong to the order Falconiformes they have little in common with the free-soaring falcons of steppe and prairie. Not only are they active in different habitats; they are strongly contrasted in appearance and behaviour. These forest raptors have short, broad and rounded wings with a large bastard wing or alula, and a long, highly mobile tail—attributes conducive not only to rapid flight but also to the sudden turns and twists demanded in their chosen surroundings.

The twilight hunter

Shortly before nightfall the metallic cries of the blackbirds echo through the woods, calling a halt to the day's activities. Crows with bills discoloured by continual pecking at the soil and

Facing page : The European sparrowhawk incubates her eggs and rears her young on a platform of dry branches in the fork of a tree. As with most birds of prey it is the female who tears up the food for her young.

Geographical distribution of the goshawk.

GOSHAWK
(*Accipiter gentilis*)

Class: Aves
Order: Falconiformes
Family: Accipitridae
Length: $19\frac{1}{4}$–$24\frac{1}{2}$ inches (48–61 cm)
Wing-length: 12–$15\frac{1}{2}$ inches (30–38·5 cm)
Wingspan: male $37\frac{1}{4}$–$40\frac{1}{2}$ inches (93–101 cm)
female 44–$47\frac{1}{4}$ inches (110–118 cm)
Diet: flesh, mainly medium-sized birds and mammals
Number of eggs: 3–4
Incubation: 35–40 days
Longevity: up to 30 years

Dark greyish-brown back; white mark above and behind eyes. Underparts white with chestnut stripes on the throat and narrow, almost black bands on breast and belly. The female is larger than the male and her back is greyer. The bill is black, the iris orange, the feet and cere yellow.

crops distended with seeds and insects come to rest in the pines and oaks. Wood pigeons circle warily above the treetops before landing on the branches with a noisy flapping of wings. Jays, choughs, redwings and other species return to their roosts before the sun sinks over the horizon.

It is precisely at this twilight hour, as the woodland birds settle down to sleep, that the goshawk makes its marauding foray. It may come gliding in like a shadow at the level of shrubs and bushes, then suddenly accelerate and veer upwards to snatch a drowsy, unsuspecting bird from a branch. Alternatively, it will allow itself to be carried along by the wind, hardly moving its short, strong wings, and swoop down like an arrow on a colony of crows, scattering them in a cloud of black feathers and vanishing into the forest gloom with a captive clutched firmly in its talons.

The raptor may well have laid its plans in advance, taking up its position in ambush during the afternoon. It perches motionless in the fork of a tree, patiently awaiting the arrival of its victim. The camouflage is perfect. Silvery, black-striped underparts, greyish-brown back and white marks over the eyes break up the silhouette of the bird in such a way that it is almost impossible to see it squatting against the trunk in the shadows, with not a movement to betray its presence.

Should one be fortunate enough to spot the bird at close quarters, its plumage glowing in the last rays of the setting sun, there is an immediate impression of concentrated power. The orange-red eyes, alert and calculating, transfix one like tiny daggers. The broad, muscular chest hints at the hidden reserves of stamina and strength; and the long tail is clearly designed for rapid and intricate aerial manoeuvres. The shining black claws—somewhat longer than those of other members of the family—curve menacingly and are obviously formidable weapons capable of killing animals larger than the raptor itself.

Although occasional migrants may be seen circling over open plains and even coming down to land there for brief periods, the goshawk is essentially a woodland bird and its hunting and breeding grounds are invariably located in forest regions.

In autumn and winter the sedentary goshawks of the Iberian peninsula, together with migrants from northern climes, establish themselves in small pine woods fringing cultivated fields, in oak woods or in dense undergrowth close to water. In hilly and mountainous areas tree growth is of lesser importance.

In the middle of a really cold winter, when animals tend to disperse and become difficult to find, the goshawks stray much farther from customary refuges and hunting grounds. At such times they have even been sighted perching near dovecotes in the centre of an orchard.

When the cold season is over the birds return to their birthplaces and traditional breeding quarters in the deepest and most inaccessible parts of the forest. Their favourite tree seems to be the beech but failing this they select a suitable conifer, such as a larch or pine. As a last recourse they choose an oak. In some parts of Europe, notably Spain, goshawks are found in all forested areas, including mountains, in summer.

Although most at home in the forest, the goshawk occasionally leaves woodland in order to hunt. It sometimes stations itself on open ground in hilly and mountainous regions, keeping a vigilant watch on prey sheltering on nearby slopes. The bird shown here is an immature goshawk.

The branch on which the bird finally alights is always selected with considerable care for it has to serve both as a vantage point for hunting and as a roost at night. When it kills a prey, however, it flies off into the thickest part of the undergrowth to remove fur or feathers and ultimately eat the animal. An ornithologist once discovered a bush about 500 yards from a pine in which a pair of goshawks were spending the winter. Under the tangled canopy of brambles he came across a massive heap of feathers which, according to his reckoning, must have been plucked from the bodies of at least fifty magpies. Not far from this spot there was a tree stump literally carpeted with the hairs of countless rabbits and rodents.

After the breeding season the goshawk becomes increasingly silent and wary. It moves abroad as little as is really necessary, making only cursory surveys of its territory, paying brief visits to the place where it customarily bathes and spending the rest of the time perched motionless on a branch or stump. At the least disturbance it flies off, skimming the undergrowth.

In winter male and female continue to share the same territory but each will fend for itself. They venture out independently in quest of food and sleep in different places, sometimes quite far apart. No matter what the season they will not tolerate intruders and will defend their territorial frontiers with great

The goshawk, sometimes used in falconry, is a low-level hunter. Although it is adept in killing birds its principal victims are mammals. When they were plentiful it specialised in catching rabbits.

Facing page: Contrary to what is popularly alleged, neither the North American species of goshawk (*above*) nor the Eurasian species (*below*) habitually attack animals hunted by man, and should they do so the victims are generally sick or injured individuals. In this respect the raptors exercise a desirable control on the populations of such species.

determination. Fierce fights may take place between rival claimants, sometimes ending in death.

The goshawk will resort to one of several flight methods, depending on the circumstances. Sometimes, as during migration journeys, it will cruise at high altitude, in which case the wings are not fully deployed. A series of strong, rapid wing beats will be followed by long glides in which the opened tail is used as a rudder and direction changer. This alternating rhythm makes for considerable speed.

The characteristic hunting flight is far more acrobatic. From its starting point on a branch the goshawk accelerates rapidly by flapping the wings vigorously and reaches maximum speed after covering only a few yards, threading its way through trees and undergrowth with incredible expertise, avoiding obstacles by a hairsbreadth. Once within striking distance of its prey several options are open to the raptor, depending on the amount of space available. The most direct course is to grab a bird in full flight, but it may just as easily outflank its victim, swivel round and attack from the front, seizing it by the head. Alternative techniques include flying alongside, making a right-angle turn and coming at the prey sideways; soaring up from below and ripping open the belly; or diving down like a peregrine falcon.

The goshawk is highly accomplished at gliding and this form of flight is especially common in the breeding season. The courtship display cannot be mistaken for that of any other raptor. The rounded wings and fan-shaped tail are outlined clearly as the bird climbs in narrow spirals. Having gained altitude it allows itself to be supported on thermal currents, gliding around lazily before plummeting towards the treetops.

This versatility is not always appreciated by falconers, despite the fact that the latter often employ goshawks for low-level hunting. A wild goshawk is liable to treat a trained bird in much the same way as any other potential victim.

Death among the trees

A predatory mammal may go to great lengths to familiarise itself with every feature of the locality in which it can reasonably expect to discover a victim. By exploring all the paths and tracks inside its hunting territory it will learn to recognise the fallow fields where partridges collect, the thickets that conceal hares, the roosts and drinking places of wood pigeons, the bushes in which blackbirds gather, the hiding places of magpies and the like. Yet it is a mere amateur compared with an experienced goshawk keeping watch from dawn to dusk and allowing nothing to escape its notice. The most insignificant happenings are etched on its memory. It knows by instinct exactly which branch to perch on so as to intercept a bird bound for its nest, which hedge to hide behind in order to capture a small magpie and the precise time when the wood pigeons are likely to venture down to the water to drink. The goshawk is concerned with the minutest activities of all the forest inhabitants for they rank either as prey or foe, and the law, here as elsewhere, is eat or be eaten.

No bird of prey is more adept in the technique of ambush. As it

The goshawk is a versatile hunter. It may, for example, perch in ambush on a bare branch and glide silently in pursuit of a passing bird, dipping towards the forest floor and veering upwards again to strike at its victim from below. If it spots a rabbit on the ground it will zigzag through the bushes until it is in a position to launch a surprise attack.

perches high on a bare branch it is unremittingly vigilant, making fullest use of its keen eyesight to scrutinise the movements not only of animals close to hand but also some distance away. Alternatively, it will take off from a hilltop and head for the wooded valley, skimming the undergrowth without moving its wings, weaving an unerring path through the ground vegetation and suddenly swooping from behind a bush to seize a victim in its vice-like grip. The sharp claws rip the flesh and death is instantaneous as they penetrate brain or heart. Now the meal can begin, although the raptor does not touch the prey with its beak until every sign of life has gone.

If the dead animal is large the goshawk drags it along the ground by stages, stripping off a piece of flesh at intervals until it is light enough to be carried away to a safe place where fur or feathers can be removed and the carcase torn up at leisure. Should the victim be a hare, an adult rabbit or a pheasant–too big to be consumed in one sitting–the goshawk will eat as much of it as possible and then conceal the remains as best it can near a bush or in the low fork of a tree. Later it retires to a neighbouring tree to stand guard over the prize and woe betide any crow or kite rash enough to attempt to filch it. Provided no fox turns up to claim a free meal there will be sufficient food for the following day and perhaps for several more, in which case the raptor need not go hunting for a while.

Whether the goshawk uses its favourite method of ambush and surprise attack or some other strategy to suit the circumstances, it seldom fails in its objective, whether it be a mouse, a hare, a thrush or a grouse. The zigzagging flight of the raptor spells almost certain death. The only possible chance of evading the predator's clutches is to remain absolutely quiet and still, hopeful of avoiding detection. The slightest sound or movement will immediately alert the raptor and unleash an attack. In the view of Felix Rodriguez de la Fuente, a trained falcon can flush out a partridge or other bird from a bush even if there is no sign of life; but a goshawk would probably not find it.

The hunting grounds of a sedentary pair of goshawks may range from 5,000–10,000 acres, but where game is plentiful the extent of the territory will be smaller, and vice-versa.

Courtship and reproduction

In Mediterranean countries the goshawks' breeding season varies from place to place; but it generally occurs somewhat earlier than in northern regions. Thus the birds may begin to construct an eyrie or renovate an old nest early in February, commencing their courtship ritual at the same time. Normally tranquil and silent, they suddenly become extremely noisy and excited.

From dawn onwards the forest resounds to their piercing war cries and soft, plaintive love calls. The female's voice seems to be harsher than that of her mate. Perched on high branches some distance from the eyrie, the birds perform curious bobbing movements and then chase each other through the trees. Sometimes they soar to a considerable height, gliding in slow circles, diving at high speed until they almost touch the trees and veering upwards to repeat the performance time and time again.

During the breeding season the male hunts more frequently than usual and brings food offerings to his partner, as well as collecting branches and twigs for the nest. From that time until the end of March or mid-April the birds proceed to couple vigorously and frequently. Juan Vallés, a 16th-century falconer, claimed that goshawks would mate forty or fifty times a day, with as much haste and passion as sparrows.

Building the eyrie

Goshawks usually choose the most dense and impenetrable parts of the forest as a nesting site. The ideal location for the eyrie is in the fork of a tree, provided the supporting branch is stout, secure and well sheltered from the elements. In a conifer the nest is often constructed in the highest branches but if, for example, a beech is selected the eyrie will be built lower down since the upper branches of this species may be too thin and fragile to bear the weight.

The dimensions of the eyrie vary, depending on whether it is a new construction or a renovated one which has already seen several years' service. It may be fairly modest in size or an enormous, complicated platform of branches piled more than 3 feet high and measuring 4–5 feet across. Whatever the size, the nest will be lined with soft vegetation carefully collected either from the ground or from trees (not necessarily material which is most conveniently available close by); and whatever the foundation of the nest, the finishing touches generally include a

The goshawk's eyrie may be a large construction of branches and twigs, lined with pine needles, and situated in the fork of a tree. If a larch or pine is selected for the purpose, the nest will be near the top of the tree. It will be used for successive seasons and enlarged as necessary.

In the course of its courtship display the male goshawk ruffles his under tail-coverts (1), then spreads his wings and tail to rise on warm air currents (2), finally falling in a high-speed dive.

Facing page : Young goshawks are extremely voracious and have to be fed by the mother several times a day. Growth is rapid and within a month they are able to rip apart prey provided for them by the parents. This coincides with the time of their first flights.

layer of pine needles. An ornithologist found one goshawk's nest in an arbutus which had been lined with oak leaves. The lining is kept fresh and periodically renewed.

Occasionally the nesting site will be close to a clearing rather than in the heart of the forest; nor is it uncommon for a pair of goshawks to prepare two or three eyries some distance apart, making a final choice at the last possible moment. If the opportunity offers they may take over a buzzard's nest—a practice which the latter just as often reciprocates.

Egg-laying and incubation

Some time between the end of March and the middle of April the female goshawk lays up to four eggs (the number is three when the breeding birds are fully mature) at the rate of one every two to four days. The eggs are light green and a little larger than those of a domestic hen.

Incubation lasts 35–40 days, during which period the female seldom budges either at night or by day. She will not even move if people happen to pass directly below the tree, but she may fly off if there is a deliberate attempt to frighten or excite her. Meanwhile the male brings her food, although he prudently refrains from venturing too close to the eyrie, calling her from a nearby tree. She will then fly over to collect the prey. The only other time she may temporarily leave the nest is to bathe. During such absences the male takes her place.

Rearing the fledglings

The eggs do not hatch simultaneously and the nestlings may thus be hatched several hours or even a day apart. For the first four or five days the mother covers them with her body, especially when the weather is cold. From that time on the male perches on the edge of the eyrie when he brings food to the family and the female helps him to pluck the feathers and dismember the prey for her offspring.

The newly hatched chicks are covered with white down, their eyes are blue-grey and the tarsi are yellow. They grow very rapidly but for the first fortnight lie on their comfortable bed of leaves, legs folded underneath the body, heads well down out of view. Like other members of the Accipitridae they are already capable of directing their excrement out of the nest to be deposited in long white streaks on the ground below. This is in keeping with the astonishingly clean habits of the adult birds—all the more surprising when one considers how many different kinds of animal are consumed inside the nest every day. No scrap of food or waste matter is allowed to foul the interior or outer surface of the eyrie. The female assiduously carries away all solid left-overs, together with twigs and leaves that fall on the nest, and deposits them at a safe distance from the site.

At the age of about two weeks the first feathers begin to appear although the thick downy covering will not be entirely shed until the youngsters are old enough to break the parental bonds. Clad in this mixed livery of down and feathers the fledglings

Winter is a challenging season for the goshawk which must extend its hunting grounds to seek prey as it becomes more scarce. The bird survives by making best possible use of its natural speed and agility, coupled with the strength of its claws and short, curved bill.

now show unmistakable signs of increasing activity though at first they simply hover around the edges of the eyrie flapping their wings from time to time. They are by this time able to eat unaided, the mother merely placing the bodies of fully feathered birds within their reach. The youngsters' claws are already sharp and they have little difficulty in ripping the toughest carcasses to shreds. Soon they start exploring the nearby branches. Their eyes gradually change to greenish-yellow, as do feet and cere.

Some forty days after birth the young goshawks are practising their flying skills and perching confidently on trees a little distance from the eyrie. The mother still spends most of her time supervising their activities and if necessary repelling intruders, whether they be mammals or birds; but she lacks the determination and courage of the female peregrine falcon, particularly when facing humans.

Even when the fledglings are old enough to fly some distance from the nest they continue returning to it regularly to eat and sleep, signalling their presence to the adults by emitting shrill cries.

In the event of the mother dying before her chicks are reared, the male allows them to die for he is incapable of taking her place; yet should this happen when the youngsters are three weeks old or more he appears to be able to give them satisfactory care and attention.

Training for adulthood

Training for adult responsibilities usually begins early in July – perhaps a little sooner or later according to the region. The female is the more active of the two parents in teaching the youngsters how to hunt, initially placing live prey within their reach so that they can practise the techniques of trapping and killing. Step by step they become more confident and agile, and after a week or so their aim is accurate and deadly. The adults do not at this stage dispute the prey, permitting the youngsters to reap the reward of their efforts.

By August the young birds, rapidly gaining strength, are ready to roam their father's hunting grounds to find their own prey, although they still return at intervals to the eyrie. But in early autumn they finally abandon their birthplace and wander off to fend for themselves, as do most other birds of prey at this stage of life.

The seasonal dispersal of immature goshawks is to a great extent the responsibility of the adults who now begin to treat their offspring as if they were complete strangers and potential enemies. One case has been reported of a mother attacking her youngster, leaving it badly injured on the ground and eventually forcing it to fly away.

Winter sees the young goshawks leading a nomadic existence but seldom straying more than 60 miles or thereabouts from the place where they were born. When the cold season is over, prior to their first moult, these birds are capable of mating and procreating in their turn. The partner will often be a mature adult who will initiate the younger bird in nest-building.

The goshawk–friend or foe?

A great deal has been said and written concerning the harm allegedly caused by goshawks as a consequence of predatory activities against game birds and mammals–enough at any rate for the raptors to be castigated in some quarters as pests high on the list of man's natural enemies. The accusations are utterly without foundation, arising from slanderous statements by certain 19th-century naturalists who accepted the unverified reports of simple countryfolk and hunters rather than rely on their own observations and experience. The evil reputation of the goshawk has thus been exaggerated to such a degree that it is popularly credited with the capacity for consuming impossibly large quantities of partridges, pheasants, hares and rabbits.

Serious surveys by modern ornithologists, scrupulously trying to be objective, have since revealed that the goshawk, far from being a harmful bird, is in fact a beneficial species.

In one characteristic field study designed to test this claim, a naturalist took up his position in a hide about 15 feet from an eyrie occupied by a pair of goshawks and noted that most of the animals brought back to the nest were jays and magpies, with a few lizards and mice. Of five partridges counted, four were spurred males; and the half dozen young rabbits that had been captured came from a litter whose mother was an obvious victim of myxomatosis.

Goshawks effectively control the numbers of forest-dwelling Corvidae–prolific birds which because of their numbers do real damage by stealing eggs and killing fledglings of many species, as well as small and medium-sized mammals. In cases where goshawks hunt game animals they–like other predators–usually attack weak and handicapped individuals so that in the long run they play a stock-improving role. To take another example, by killing male partridges that are still in rut while the females are incubating the eggs, the goshawks indirectly help to assure the survival of the colony.

Dr Brüll and his colleagues conducted a statistical survey in Germany between 1950 and 1955, keeping watch on a pair of goshawks patrolling hunting grounds that covered some 10,000 acres of woodland. The naturalists reported that the birds killed 1,395 animals belonging to some 60 different species during this period. Among the mammals were rabbits, hares, leverets, rats and water voles, as well as a number of predators such as polecats and weasels. Bird prey included a large number of crows, wood pigeons, domestic pigeons, magpies and jays as well as some partridges, starlings and other small and medium-sized species.

The ornithologist Otto Uttendörfer, one of the first scientists to make a serious study of the habits of the goshawk, examined the contents of 245 eyries and meticulously listed 7,333 animals of which 647 were mammals (most of them rodents) and 6,686 birds (including 1,153 jays, 319 crows, 96 magpies, 12 ravens and 11 jackdaws).

In Spain, before the myxomatosis epidemic, the normal diet of the goshawk population consisted in the main of rabbits, supplemented during the winter by wood pigeons and all the year round

The young goshawk's plumage changes colour at about two years of age. Prior to the first moult the feathers are varying shades of brown, without striking contrasts. When the plumage takes on its permanent adult appearance the back turns darker and the white underparts are conspicuously streaked with black.

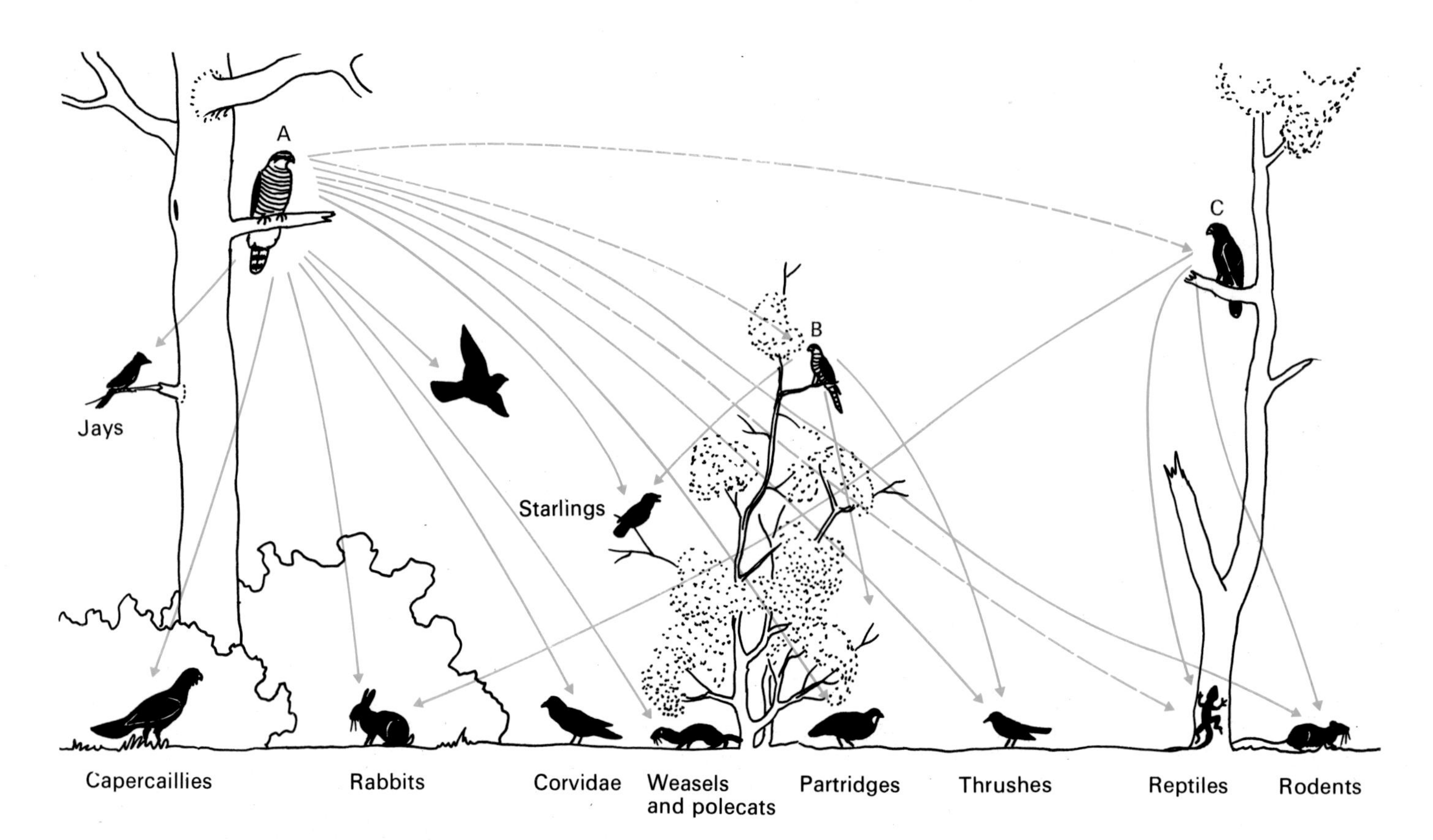

Although all three raptors nest in trees, the goshawk (A), sparrowhawk (B) and buzzard (C) do not compete directly with one another. The first two both hunt in the undergrowth but do not pursue the same types of prey. The buzzard, which is less agile, prefers to roam the open spaces. This diagram gives some idea of their respective hunting and feeding habits.

by magpies, crows and jays, especially in the breeding season when the males would pillage the nests of the Corvidae and kill the fledglings as they embarked on their first flights.

Nowadays goshawks are only found in small numbers because of the deplorable persecution to which they are still subjected. Their range is virtually confined to heavily wooded areas and mountain regions; and their presence in these parts is a sure guide to the existence of a richly varied regional fauna. Conversely, such animals would not flourish abundantly if the raptors were to vanish entirely from their midst.

There are nine distinct subspecies of goshawk in the northern hemisphere. The bird is found almost throughout the Palearctic region as far south as Morocco and as far east as Iran, Tibet, Siberia and Japan. In America the raptor is an inhabitant of most of the large Canadian and Alaskan forests, ranging southwards into California and northern Mexico.

The diminutive sparrowhawk

By the side of the goshawk, another characteristic raptor of the forests–the sparrowhawk–seems little more than a dwarf. But despite its moderate size it is also a formidable predator, specialising in hunting other birds. It has an elegant body and its long, slender claws are reminiscent of the talons of falcons. The Eurasian species nests in woods but frequents coppices and thickets close to rivers and streams during the winter. In fact the bird will settle wherever it can find sheltering shrubs and trees and is often seen in orchards and public gardens.

The sparrowhawk is arguably even more agile and acrobatic than the goshawk; certainly no other bird of prey can compare

Facing page : Cooper's sparrowhawk, a resident of the North American forests, is similar in colour to the European sparrowhawk and the goshawk, but is larger than the former and smaller than the latter.

with it for rapid acceleration. Taking off from a low branch, it can reach maximum velocity with only a few vigorous wing beats and maintain this speed for 200–300 yards. Attacks are carried out to the accompaniment of complex aerial manoeuvres and few victims have sufficiently fast reactions to avoid capture.

More prolific than the goshawk, the female sparrowhawk lays 3–6 eggs and incubates them for 36 days. The fledglings leave the eyrie very early and after a short period of apprenticeship fly off to roam the territory previously monopolised by their father. Those birds born in northern lands depart on migration journeys, as do goshawks, many of them crossing the Mediterranean, but a large proportion spending the winter in the Iberian peninsula.

No European bird of prey has suffered such a rapid and sad decline in numbers as the sparrowhawk, partly as a result of hunting but mainly as a consequence of ingesting the poisonous insecticides that have accumulated in the body tissues of their victims. When this does not result in death the probability is that the bird will become sterile.

In North America the handsome Cooper's sparrowhawk (*Accipiter cooperi*) is not only midway in size between the goshawk and the European sparrowhawk but also combines the anatomical features of both its relatives. The body is more slender and streamlined than that of the goshawk but not as graceful as that of the common sparrowhawk. This raptor has a broadly based diet but kills many more birds than mammals.

The female of *Accipiter nisus* has plumage which is similar in hue to that of the goshawk while the male has a rufous breast and belly. In *Accipiter cooperi* these brighter colours are seen in both sexes though they are more conspicuous in the male.

The various representatives of the Accipitridae of the Eurasian and North American forests in fact vary greatly in size and weight. The male European sparrowhawk, for example, which is smaller than the female, weighs a mere $\frac{1}{4}$ lb, its American counterpart about $1\frac{1}{2}$ lb and the female of the northern subspecies of goshawk approximately $3\frac{1}{2}$ lb. All these raptors are capable of killing prey considerably larger than themselves and thus account for an extremely varied range of victims.

The buzzard

The buzzard (*Buteo buteo*) is easily the most common European bird of prey but except during the breeding season its habitat is not restricted to forests. Following this period it abandons the shelter of woods and dense undergrowth in favour of spinneys, fields and even open steppe, where large numbers of rodents roam.

The rows of black poplar and willow that line rivers and streams are favourite vantage points for the buzzard, but failing these a telegraph pole or picket fence will serve just as well.

In the depths of winter, when the bare branches and menacing grey skies are enough to convince the casual observer that plant and animal life are dead, the endless struggle for survival is in fact at its grimmest, with only the fittest, best endowed individuals emerging unscathed. But it requires an experienced naturalist, prepared to brave the rigours of the climate, to appreciate the

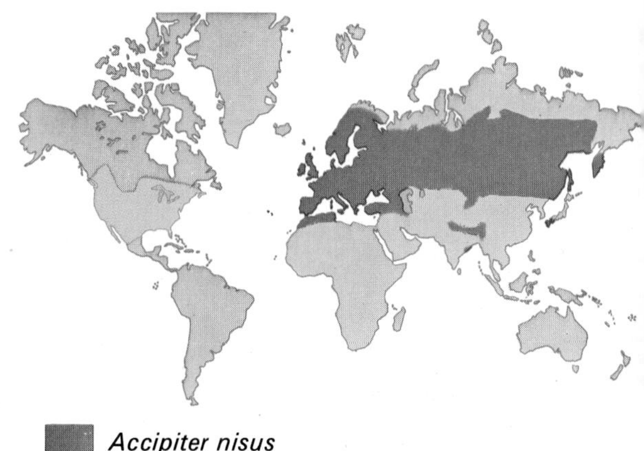

Geographical distribution of the European sparrowhawk (*Accipiter nisus*) and Cooper's sparrowhawk (*Accipiter cooperi*).

SPARROWHAWK
(*Accipiter nisus*)

Class: Aves
Order: Falconiformes
Family: Accipitridae
Length: $12\frac{1}{2}$–$15\frac{1}{4}$ inches (31–38 cm)
Wing-length: $7\frac{1}{2}$–$9\frac{1}{2}$ inches (19–24 cm)
Wingspan: male $23\frac{1}{2}$–26 inches (59–65 cm)
female $27\frac{1}{4}$–$30\frac{3}{4}$ inches (68–77 cm)
Diet: flesh, principally birds
Number of eggs: 3–6
Incubation: 36 days
Longevity: up to 12 years

Similar to the goshawk but smaller. Yellow iris. Back blue-grey; belly of male rufous, that of female white with grey stripes. Young bird has chestnut-striped back.

COOPER'S SPARROWHAWK
(*Accipiter cooperi*)

Class: Aves
Order: Falconiformes
Family: Accipitridae
Length: 14–20 inches (35–50 cm)
Diet: flesh, more birds than mammals
Number of eggs: 3–6
Incubation: 36 days

Size between that of goshawk and European sparrowhawk. No white eye-marks. Back blue-grey, belly fawn, striped with darker brown; male's back darker and underparts lighter than female. Iris orange.

Facing page : Because of intensive hunting and the indirect effects of chemical pesticides it is increasingly rare in Europe to come across an eyrie of a sparrowhawk in a tree (*above*) or to see a buzzard in a clearing (*below*).

Geographical distribution of the buzzard.

BUZZARD
(*Buteo buteo*)

Class: Aves
Order: Falconiformes
Family: Accipitridae
Length: 20½–22½ inches (51–56 cm)
Wing-length: 14¾ inches (38 cm)
Wingspan: 48–56 inches (120–140 cm)
Weight: male 1½–2 lb (600–900 g)
female 1¾–3 lb (800–1200 g)
Diet: flesh, mainly rodents, at one time chiefly rabbits
Number of eggs: 1–3
Incubation: 33–35 days
Longevity: up to 25 years

Adults
Much individual variation. Underside of wings usually white. Broad, rounded tail; small neck. Plumage generally dark brown with white stripes on back; breast and belly pale, streaked pink or black.

Young
Similar to adults.

Facing page: Although the buzzard is now primarily a hunter of rodents it will, should the opportunity occur, capture a currently rare rabbit as well.

finer points of this life-and-death fight against the cold; and the ornithologist watching patiently from the comparative comfort of his hide may find ample reward in the spectacle of a buzzard which, like all other predators, cannot afford to relax its hunting activities in its need to survive.

The naturalist's first sight of the raptor may be a small, vaguely defined, pinkish shape at the top of a poplar, the branches of which tremble in the wind. Head buried deep between the shoulders, feathers ruffled, one foot concealed in the warm down of the belly, the buzzard faces the icy northern blasts without moving a muscle. Although there is nothing in this huddled ball to indicate the tension of a raptor in ambush, the bright amber eyes of the bird are constantly on the alert. Its gaze alights on a row of brown moorhens as they move slowly forward on dainty green legs, searching for earthworms; but well aware that it is a waste of time to attack these agile aquatic birds, the buzzard's interest soon lapses. A group of ducks likewise fail to engage the raptor's attention, and the twittering band of finches pecking around the base of the tree are equally safe.

Suddenly the buzzard stiffens. The foot which until now has been hidden in the plumage is abruptly extended, the feathers are flattened against the body, the aquiline profile of the head is clearly visible, the flashing eyes are intent and unflinching. Close scrutiny soon reveals the object of the bird's interest—a vole which has popped out of the undergrowth alongside a snowy path already laced with tiny footprints. Its destination is a clump of trees on the opposite side of the field, but before setting off to sample the tasty bark the little rodent makes a careful survey of its surroundings. But its eyesight is so poor that it fails to spot the lurking buzzard which is still perching motionless on the poplar high above. Suddenly it scuttles into the open. When it is about half-way across the field, the buzzard finally makes a move. With a few slow, powerful wing beats its glides effortlessly down towards its victim. The vole makes a desperate attempt to retrace its footsteps but it is too late. A few yards from the safety of the undergrowth it is seized in the raptor's claws and killed instantaneously.

This is the characteristic hunting technique of a bird which specialises in preying on small and medium-sized mammals; but the buzzard will also patrol its territory in quest of such victims, using its broad wings and rounded tail for protracted gliding flights. It may hover for hours above fields and forest glades but seldom lingers in regions where tall grass and clustered trees would only hamper its activities.

The buzzard feeds principally on mice, voles, shrews, moles and (formerly) rabbits, but will not disdain insects, for examination of stomach contents have revealed large quantities of beetles, crickets and grasshoppers. In hot weather it hunts lizards and snakes, and when particularly hungry it will consume carrion. Because of its size it is seldom successful in attacks on mammals larger than rabbits and it hunts especially around burrows where rabbits are suffering from myxomatosis. Here is further evidence that sick animals tend to be methodically eliminated in regions where the balance of nature is still relatively stable.

It should be evident that the buzzard is an extremely valuable bird of prey because of the efficient manner in which it disposes of vermin. The peasants of the Ukraine, recognising this fact, place special roosts in the fields so that the raptors can perch on them and keep constant watch for rodents.

As far back as 1902 the Spanish government led the way by promulgating laws designed to protect these birds. These regulations, though still in force, are unfortunately not strictly applied. Furthermore, it is sad to remark that members of hunting organisations in some countries as well as the editors of certain prominent hunting journals repeatedly stress the need to get rid of these birds at all costs. It is such misguided individuals who must bear the responsibility for the serious proliferation of rats and mice in many agricultural areas.

During February, as the days imperceptibly grow longer and the countryside slowly awakens from its winter torpor, the buzzards give vocal evidence of their intention to mate. They utter piercing, rather cat-like cries as they soar, swoop and glide in spirals for hours on end; or they pursue each other indefatigably high above the fields and woods. Such aerial displays, proof of the birds' excitability, are designed not only to entice and attract a partner for breeding purposes but also to mark out the borders of the nuptial territory.

A few weeks later, in March, the two buzzards, having paired off, build their eyrie high in the fork of a tree. Both birds carry small branches and twigs to the site, heaping them into a solid platform and lining the interior with scraps of vegetation. The tree in which this elaborate construction is prepared may be situated in the heart of a forest or stand on its own in the centre of an open plain. Alternatively, the birds will take over the used nest of a crow or raven. When all the building preparations are completed the buzzards mate.

The female lays two or three eggs around the beginning of April and proceeds to incubate them for 33–35 days. During this entire period the deep silence round the eyrie is broken only by the occasional shrill cries of the father-to-be as he brings food for his mate, this being the only time when she briefly interrupts the incubation. When the chicks are hatched she stays near them for several days to protect them and keep them warm. About a week after they are born she once more resumes her hunting activities. Soon the first feathers appear and at three weeks or thereabouts the youngsters can stand upright. When they are six weeks old they are chirping lustily to attract the attention of their parents as the latter fly close to the nest. The last chick to be born, inevitably the weakest of the brood, may, however, be pecked to death by its siblings.

The young birds begin flapping their wings at the age of one month in preparation for their first flight a couple of weeks later. Towards mid-June they leave the nest but the parents then subject them to a long two-month training course in hunting. When the immature birds are able to feed and defend themselves properly they stray far from their birthplace to lead a nomadic life until the time comes for them to find a mate and establish their own territory.

The buzzard is still fairly abundant in parts of Europe, especially the Mediterranean countries. The colour of the plumage is extremely variable, ranging from brown to almost pure white but generally blended.

Buzzards in the wild seldom reach an advanced age for they have to face virtually insurmountable problems. Though well equipped to combat their natural enemies they have no way of resisting man and the gun. Their average lifespan is no more than six or seven years although two ringed individuals aged twenty-four and twenty-five years have been recovered.

The buzzards of the Iberian peninsula do not migrate in the winter and are joined by birds that have journeyed from central Europe. Each bird sleeps and hunts within its own territory but life at this season is difficult. Were it not for their broad range of food the raptors would die of starvation and in some cases they are obliged to abandon their traditional habitats and fly off to warmer climes.

The buzzard's distribution range is vast, covering almost the entire European continent and central Asia with the exception of the most northerly regions. Despite the fact that they are the object of illegal and unrestrained hunting in Europe they are still fairly numerous in certain areas, especially in Mediterranean countries. Their Asiatic distribution extends eastwards as far as Siberia and Japan; and from the European mainland they range westwards to the Canary Islands, the Azores and the Cape Verde Islands. Over this wide area ornithologists have identified four different subspecies.

Buzzards hunt a broad range of prey especially during the breeding season and when they are rearing their offspring. Reptiles are abundant when the chicks hatch and the latter consume these in large quantities, though always rejecting the skin.

CHAPTER 12

The silent hunters of the night

As the last dying rays of the setting sun paint the sky crimson and twilight shadows soften the outlines of the hills, crested with oak and beech, the forest is enveloped in sudden silence, broken only by the occasional high-pitched cry of a blackbird. As dusk falls life seems temporarily to have been suspended. In the cool evening air birds that have been active all day settle down among the branches for the night, not stirring until the following morning.

The general impression of stillness and inactivity is of course illusory. True, the diurnal inhabitants of the forest are now at rest but the nocturnal species are just beginning to bestir themselves. Dormouse, shrew, fieldmouse and rabbit start to prepare for their nightly hunting forays. Familiar with the lie of the land these small mammals move effortlessly through trees and undergrowth without needing to use their eyes. But although they are safe from the attentions of carnivores and birds of prey that only roam by day they are continuously exposed to the attacks of a huge army of nocturnal predators ready to take the places of those that have sunk into drowsy inactivity.

As the last light vanishes and darkness shrouds the forest, the owls–most specialised of all birds of prey–open their enormous round eyes and take up their positions on the branches of elms bordering a stream, on pollarded willows close to cultivated land, on rocky outcrops in hill country or on ruined buildings. The eagle owl launches its sinister cry from a vantage point near its eyrie and the trembling call of the tawny owl echoes from the foliage. The soft, fluty song of the little owl carries across the fields and the barn owl screeches from the top of an old clock tower or steeple.

With a reputation enhanced by legend, few birds are so

Facing page : At the time when the diurnal hunters prepare for sleep the nocturnal birds of prey take their place, keeping up continuous pressure on the forest inhabitants. Like the daytime predators, the owls perform a service to farmers by capturing hordes of rodents and for this reason alone deserve the fullest protection.

From a sheltering hollow of an oak a little owl shows itself fearlessly in broad daylight. But although it is one of the few owls to venture out occasionally by day it is essentially a nocturnal hunter.

maligned and misunderstood by the public as the owls. Popular lore has long associated them with doom and disaster. Creatures of the night, with a strangely menacing appearance and terrifying voice–how could they be anything but evil? Superstition of this kind was commonplace centuries ago in remote communities of simple, uneducated people, but it is hard to understand how such beliefs could have lingered on in our supposedly more enlightened age. Yet so it is. Some people cannot see or hear an owl without an involuntary shudder of foreboding and no neat, logical scientific explanation will entirely dispel their misgivings.

The serious naturalist cannot be influenced by legends which patently distort known facts. He recognises that nocturnal activity does not imply possession by the devil and that glittering eye, hooked beak, curving claw and strident cry, far from having sinister implications, are physical attributes which have evolved under the sheer pressure of survival.

The use of the term 'raptor' to describe these birds is perhaps confusing in that it implies a close relationship with eagles, falcons and other species with diurnal habits. But careful study of the two groups reveals notable differences so that ornithologists have accepted separate classification whereby the birds of prey that are active during the day are grouped together as Falconiformes and the so-called nocturnal hunters as Strigiformes, from the Latin *strix*, meaning a screech owl.

The owls, which make up the latter group, are in fact more closely related to nightjars than to eagles, the hooked beak and sharp, curved claws being convergent adaptations to common hunting techniques. Yet to describe owls as 'nocturnal' is also slightly misleading, for some of them are active by day, notably the hawk owl of birch woods and coniferous forests, the snowy owl of the Arctic wastes and, to a lesser degree, the pygmy owl, smallest of European species. In fact, with the possible exception of the tawny owl, all representatives of the order Strigiformes are able to see reasonably well in dim light.

Nevertheless, the majority of owls prefer to hunt by night for they can distinguish small details much better in darkness and are adapted in various ways to a life spent predominantly among the woodland shadows.

Camouflage and disguise

There was a time when hunters attempted to catch eagles, falcons, goshawks and buzzards by using an eagle owl as a decoy, placing it inside a cage in the middle of a field and waiting for one of the diurnal birds of prey to swoop down and attack the unfortunate prisoner. Such methods, happily prohibited almost everywhere nowadays, were based on the recognised antipathy of the birds concerned. In the course of a survey on the food habits of the eagle owl the ornithologist Uttendörfer was able to identify remains of 15 peregrine falcons, 9 goshawks and one fish eagle; conversely, he discovered parts of dead owls in a goshawk's eyrie.

The cryptic coloration of owls' plumage is in itself a defensive weapon which helps to disguise these birds during the day,

Facing page: Two of the commonest European owls are the barn owl (*above*), easily recognised by its heart-shaped facial disc, which wraps its wings round itself when it settles down to sleep either on a branch or in a barn, and the little owl (*below*), which always seems to be wide awake.

whether they happen to be concealed amidst foliage or among rocks, for there is no single colour which stands out in contrast to any other. The pure white plumage of the snowy owl similarly shows perfect adaptation to the Arctic landscape.

Some species have a pair of stiff, horn-like ear-tufts which likewise serve to disguise the bird by breaking up the shape and outline of the head which might otherwise be too conspicuous against a particular type of background. These tufts are sometimes referred to as 'ears' but this is misleading, for they have nothing whatsoever to do with hearing. In Europe and Asia they are distinguishing features of the eagle owl, the long-eared owl and the Scops owl, and in North America they help to identify the great horned owl and the screech owl.

Darkness and death

All the owls that habitually hunt by night (the vast majority) adopt more or less the same methods. Leaving their daytime roosts as the evening shadows lengthen they take up strategic positions inside their territory and emit characteristic hoots which are often echoed by their companions close by.

It was long believed that owls managed to find their way in the darkness by means of sight alone. But although the retina of an owl's eye is sensitive enough to enable the bird to spot a mouse on the forest floor some distance away, even on a fairly dark night, ornithologists have now come to the conclusion that an owl is not primarily guided by vision as it flits silently and unerringly from branch to branch in the gathering gloom. In fact, experience will already have familiarised the bird with every last detail of its hunting terrain so that it knows precisely the position and shape of each branch, rock and tree trunk within its territory. Such experience is gained by repeated visits to a particular locality so that all natural obstacles can be pinpointed and committed to memory for later reference.

These fact-finding surveys commence very early in life as soon as the immature bird leaves the nest for the first time, tentatively moving up and down the branch on which it is perched. As it grows older the field of action is gradually extended but once having explored an area thoroughly nothing can induce the bird to venture farther into unknown territory. Later, when it is fully adult and has established the boundaries of its hunting grounds, it will never overstep these frontiers, not even to the extent of flying across to a tree only a few yards beyond its domain, for this would entail entering hitherto unvisited and uncharted terrain and the risks of operating in such alien surroundings would be too great.

As the night passes and dawn approaches the owls which have previously proclaimed their territorial rights so noisily fall silent for the hour has now arrived for them to set out on their hunting expeditions. From their lofty vantage points they carefully scrutinise the surroundings, slowly turning their heads whenever they hear a suspicious sound nearby – the tiny footsteps of a mouse, the crackle of a dead leaf or a bird fluttering about in the branches – so that they can locate the exact spot where

prey can eventually be found. They rely entirely on their keen hearing to provide them with this vital information.

The external ear of an owl is very highly developed – in contrast to that of other birds – though the structure varies according to the species concerned. The auricular slit, frequently lunate (crescent-shaped) is bordered by two mobile folds of skin which can be opened or closed at will. Sounds are thus conveyed from the outside down a wide duct into the large, complex inner ear. The asymmetrical position of these folds in certain species means that sounds do not always reach both ears simultaneously. Although the time lapse is very small it is enough for the owl to be able to decide from which direction the noise is coming and this sound locating capacity is greatly facilitated by the ease with which the bird can swivel its head on the flexible spinal column. The eagle owl, for example, is able to turn its head through an angle of 270°. This remarkable pliancy serves a dual purpose. The owl can make a prompt about-turn in response to the faintest of night sounds and by not having to move its body and scrape its talons against the bark – a noise which might immediately betray its presence to victims whose hearing is a match for its own – it gains a considerable initial advantage in its hunting.

The eyes of an owl are frontally placed whereas those of other birds are situated at the sides of the head. By swivelling the head through a wide angle a broad field of vision can be covered. Sight, however, is not nearly so well developed as hearing and the close-up vision of an owl would seem to be rather poor. Furthermore, because the eyes cannot be moved in the eye-sockets the whole head has to be revolved every time the bird changes the direction of its gaze. The owl's ability to distinguish colours is also mediocre due to the fact that the retina contains compara-

Among the common physical characteristics of the night-hunting owls are keen hearing (1), enormous eyes sunk deeply into their sockets (2), drab, inconspicuously coloured plumage which provides excellent camouflage (3), delicately fringed borders to the primary wing feathers (4) and powerful toes equipped with slender, curved claws (5).

Facing page : Some nocturnal birds of prey, such as the eagle owl (*above*) and the long-eared owl (*below*) have prominent ear-tufts which can be freely moved up and down. These stiff tufts are unconnected with the owl's hearing but play some part in disguise by obscuring the otherwise clear outlines of the head.

Preceding pages : The long-eared owl is an inhabitant of coniferous forests and will often take advantage of nests that have been abandoned by other raptors or members of the crow family. There are usually three or four chicks in a brood.

Perched on top of a fence or stake, the barn owl is able to hear the tiny sounds made by a rodent as it nibbles a blade of grass (1). Having pinpointed its victim and made sure that there is no obstacle to impede its attack the owl swoops silently down, claws outstretched to grasp the prey (2). It then carries the dead animal away and proceeds to swallow it without removing the skin (3).

tively few cones (cells sensitive to colours) in relation to rods (cells sensitive to weak light). It is probable that the owl sees objects in varying shades of grey.

Comparative examination of the structure of the sense organs seemed to provide overwhelming evidence that hearing rather than sight was the principal guide to the hunting activities of these nocturnal birds of prey; but in order to obtain positive proof the zoologists Roger Payne and William Drury of the Hatheway School of Conservation in Massachusetts conducted a series of interesting laboratory experiments. An owl was introduced into a completely dark, hermetically sealed room, the floor of which had been strewn with a layer of leaves. The scientists followed the movements of the owl by means of infra-red apparatus. Each time they let loose a mouse the raptor immediately turned its head in the rodent's direction. Since it evidently could not see anything it was clearly guided by the tiny pattering footsteps of the animal. The moment the mouse stopped moving the owl hurled itself on the prey with deadly accuracy. The experiment was repeated a number of times, always with the same result. It seemed almost certain that the raptor's sense of hearing had led it unerringly to its objective but there was still the likelihood that smell might have played a part. This possibility, however, was eliminated after an experiment in which the owl was seen to launch itself with equal precision on an inanimate object pulled along on a piece of string. Furthermore, in a final experiment, the scientists sealed the ear slits of the bird. Deprived of its hearing the owl was quite incapable of locating its prey.

Another question which springs to mind is why the victim does not hear the owl as it swoops down in the darkness. The simple answer is that the raptor makes virtually no sound as it glides through the night air because of the thick layer of loose, supple feathers that covers the entire body with the exception of bill and tail. The smooth surface of the plumage prevents any rubbing noise and in addition the external vexilla (vanes) of some of the primary wing feathers form a kind of comb, the teeth of which similarly deaden the sound of the wing beats. According to Thorpe and Griffin the flight action does not even produce telltale ultrasonic vibrations.

If it were not for these adaptations owls would certainly find it difficult if not impossible to capture prey in the forest, considering that the length of their wings and tail do not permit those quick and sudden changes of direction which are normally essential for predators trying to weave their path through the dense tangle of branches and undergrowth.

The first and most vital phase of the owl's hunting plan is therefore to locate the whereabouts of the victim by carefully listening for its movements. Then, provided it is not pitch dark, this will be followed by a visual survey of the surrounding terrain in order to pinpoint the exact position of the animal in question and to make sure that there are no obstacles in the line of projected attack. The owl then swoops silently to the kill. The powerful claws close tightly round the body of the prey and death is quick.

There is the least possible delay in consuming the dead animal

but because of shortsightedness the owl may experience some initial difficulty in focusing its vision. It may therefore lightly graze the victim with bill and vibrissae prior to eating it. If the animal is too large to be consumed in a single gulp it will be systematically dissected but smaller victims such as mice, shrews and fledglings are swallowed whole. The stomach juices, however, are incapable of digesting certain parts of animals' bodies, notably bones, hair, feathers, claws and beaks, and these have to be regurgitated, travelling the whole length of the esophagus some hours after having been ingested and before any more food can be taken. This is not as inconvenient as it sounds because the undigested scraps are compressed into compact 'pellets', with the hardest material towards the centre. This method of disposing of waste matter is not of course exclusive to owls. The Nectariniidae (sunbirds), for example, are only one of many bird families in which this phenomenon is found. Nevertheless, the Strigiformes exemplify it to a more marked degree than other groups. It provides valuable information concerning their food habits in particular areas as well as details about local fauna. By careful examination of the pellets accumulating under the nests and roosts of different owl species naturalists have been able to collect a representative sample of all the small insectivores and rodents in regions where such raptors are present. Thanks to such surveys scientists have proved conclusively that the predatory activities of owls are beneficial to agriculture, checking the excessive proliferation of harmful species that, if left uncontrolled, would play havoc with local crops.

Although many nocturnal animals have fairly good eyesight they do not necessarily make use of it when they move about in the darkness. Experience and the daily use of the same routes familiarises them with the lie of the land. Thus if all natural obstacles are removed from the path used by a mouse to scuttle back to its nest it will still continue to jump over or run round them as if they were still in its way.

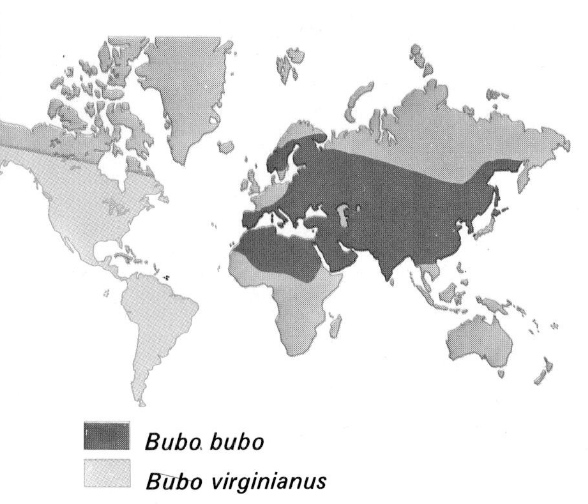

Bubo bubo

Bubo virginianus

Geographical distribution of the eagle owl (*Bubo bubo*) and the great horned owl (*Bubo virginianus*).

EAGLE OWL
(Bubo bubo)

Class: Aves
Order: Strigiformes
Family: Strigidae
Length: 26–28 inches (66–71 cm)
Wing-length: $17\frac{1}{4}$–$18\frac{1}{4}$ inches (43–46 cm)
Weight: up to 6 lb (2·7 kg)
Diet: flesh, mainly rodents
Number of eggs: 2–5
Incubation: 35 days
Longevity: 60 years in captivity

Adults
Largest of European nocturnal birds of prey. Enormous orange eyes; ear-tufts; white marking on neck only visible during nuptial display. Colour light brown with darker oval spots.

Young
Covered with white down at birth, turning grey with stripes in immature bird; adult plumage at about three years.

Only in very rare cases have pellets revealed the remains of partridges and other game birds and this is easily explained by the fact that most victims of owls are animals that are active by night. As soon as the sun sets pheasants, partridges, quails and the like prepare for their nightly rest, making no sounds which might betray them to lurking nocturnal predators. There are of course exceptions but for the most part diurnal birds are fairly safe from the attacks of owls. Owls sometimes beat with their wings at ivy and capture small birds disturbed from their roost.

The eagle owl, prince of the night

Towards the end of February, when the earliest plants begin to show shoots and the fruit trees are bursting into bloom, even though some mountain slopes are still deep in snow, the eagle owls noisily proclaim their mating desires.

The eagle owl (*Bubo bubo*) is the largest of the European Strigiformes and at this time of year the feathers adhere so closely to the body as to present an unbroken velvety brown surface. The male, head held high, neck inflated and with the white mark around the throat clearly visible, begins his courtship ritual at dusk, circling his partner and continuing the performance for much of the night. According to Hagen both birds soar high up into the evening sky, but that is about all we know of their mating habits, so rare have they become in the wild and so difficult is it to follow their activities in the gathering night shadows.

Eagle owls have adapted to climatic contrasts so successfully that they are able to settle in diverse habitats ranging from the Sahara to the tundra and from sea level to heights of 16,000 feet and more. Thanks to this extensive distribution one female may lay her eggs in a region where it is still bitterly cold at the same time as another in the sun-drenched Mediterranean basin. In Europe this is normally between mid-March and mid-April. From two to five round white eggs are laid in the nest, or more precisely on the site where the owl eventually rears its young, for the adults do not construct a proper nest, choosing either a hollow in a tree trunk, a rock ledge or a platform of twigs and branches previously prepared by a falcon, a stork or even a heron.

Incubation, which commences as soon as the first egg is laid, lasts five weeks, some of the chicks being born well before the others and preserving their size and weight advantage throughout the normal growth period. This can prove a serious liability for the youngest members of the brood when food is scarce. The adult birds, which can be relied on to defend their progeny with great courage, will not hesitate to kill them in times of severe food shortage, but the chances are that the smallest chicks will already have been disposed of by their stronger siblings. This is an effective method of controlling population growth and in no way implies deliberate, premeditated cruelty. Numbers are geared to the amount of prey likely to be obtainable. In times of plenty most paired individuals reproduce and almost all the young birds survive; but when conditions are adverse a large number of females fail to lay eggs at all and those who do are compelled to

Facing page (above): Young great horned owls move about on branches close to the nest before venturing on their first flight. (*Below*) The female eagle owl keeps watch on her progeny until they are able to stand on their own feet, then leaves them to look after themselves while she joins her partner in hunting for food.

An eagle owl will stay silent and motionless during the day so as not to draw attention to itself but if taken by surprise will do its best to frighten off the intruder. By hunching the body forwards, ruffling the feathers and half-opening the wings it can give the impression of being two or three times its normal size while a peculiar snapping of the bill helps to reinforce the alarming appearance of the bird.

sacrifice some of the brood, those that are not eliminated prior to leaving the nest dying later of starvation because they have not yet perfected their hunting techniques. Such annual fluctuations in population levels are more common in northern countries such as the Soviet Union where there is a direct relationship between the numbers of hares and eagle owls.

The male normally undertakes the task of feeding his partner while she is incubating, bringing her prey—either in his bill or between his claws, depending on the size—several times a night. He does not offer his prize directly to her but lays it down close to the nesting site in a special corner reserved for the purpose where all feeding takes place.

When morning comes the male eagle owl settles down in the shelter of a rock, on a branch or even on the ground, preferably in the middle of a bush, there to sit silent and motionless. Should it have to defend itself during these hours of enforced inactivity it makes use of an extraordinarily subtle means of dissuasion. When discovered, it leans forward, swells its feathers and half-opens its wings so that it looks two or three times its usual size. Then it glares fixedly at the enemy, making strange snapping noises with its bill. Its appearance is now so grotesque and terrifying that, if nothing else, the aggressor is taken momentarily by surprise, giving the owl enough time to escape. This menacing attitude is clearly inborn rather than acquired for the fledglings may adopt it only a few days after they are born. If it proves efficacious in the case of a solitary individual, it is all the more so when an entire brood face an intruder, ruffling their feathers, hissing angrily and snapping their bills in unison.

Food preferences of the eagle owl.

Young eagle owls stray some distance from the nesting site before they are able to fly and the parents feed them on nearby rocks. At seven weeks they attempt their first flights but it is ten weeks before they are capable of launching themselves from the top of a rock and gliding down with slow, relaxed wing beats towards the valley below.

The family remain united throughout the summer and it is during this comparatively easy and undisturbed period that the chicks learn to hunt on their own. When at the beginning of autumn the leaves turn yellow, some of them already torn from the trees by the first gusts of the north wind and littering the forest floor, the hoots of the owls once more echo through the glades at dusk. But now they do not announce the mating season and have a more sinister import, for they are adult warnings to the immature birds that the time has come to assert their independence and fly off to find their own hunting grounds. This is therefore the season when individual eagle owls may be seen setting off on long journeys which take them far beyond their habitual range of distribution–to northern France, Holland, Denmark and the British Isles.

Owls' eyes are situated in the front of the head instead of at the sides, as is the case with other birds. This gives them good binocular vision but their range is limited because of the tubular shape of the eyes. They are forced to move the head to change the direction of their gaze.

Prey of the eagle owl

The eagle owl is the most powerful of the nocturnal forest hunters and its victims may range from scarab beetles to kids and fawns. But careful scrutiny of pellets shows that the bird is a specialised killer of rodents. Out of 487 food items identified by Curry-Lindahl in Sweden, 270 consisted of mammals, including 37 squirrels and 209 other rodents, 21 hares, two mustelids and one cat. Of 158 birds there were 46 Corvidae, 41 pheasants, partridges or quails, 23 ducks, 17 gulls, nine diurnal birds of prey, four noc-

Geographical distribution of the tawny owl (*Strix aluco*) and the barred owl (*Strix varia*).

TAWNY OWL *(Strix aluco)*
Class: Aves Order: Strigiformes Family: Strigidae Length: up to 18 inches (46 cm) Wing-length: $10\frac{3}{4}$–$11\frac{3}{4}$ inches (27–29·5 cm) Weight: up to $1\frac{1}{2}$ lb (685 g) Diet: flesh, mainly rodents Number of eggs: 2–4 Incubation: 30 days
Adults Strong, sturdy appearance. Black eyes; no ear-tufts; greyish or pinkish facial discs. Colour variable–brown, grey or russet.
Young Covered with white down at birth. Plumage of immature bird part feathers, part down.

Facing page: The tawny owl flits from tree to tree inside its territory and returns to its vantage point to devour its prey. But in the breeding season it brings back such prey to the nest to satisfy the appetites of the greedy fledglings who even at an early age are able to swallow large rodents.

turnal raptors, three coots, a grebe, a woodcock, a curlew, a pigeon, a cuckoo and ten other unidentified species. The remaining items were 53 fishes, four frogs and two reptiles.

A similar experiment carried out in the Carpathians by Uttendörfer showed that out of a total of 596 animals captured, there were 314 rodents (including five squirrels), 23 hedgehogs, twelve hares, two moles, one shrew, four mustelids, ten game birds, six nocturnal raptors, seven Corvidae, a duck and 24 unidentifiable species; also 184 frogs and toads and eight fishes.

Although many other food analyses have been made elsewhere, these two alone indicate that the food of the eagle owl, despite regional variations, consists in the main of rodents–in these cases representing 51 and 52 per cent of total prey. The remainder is made up of a large variety of animals, of which only frogs appear with any regularity.

These diverse reports on the hunting and feeding habits of the eagle owl are sufficient to demonstrate the immense value of this predator, which rids the countryside of hordes of vermin. It proves that the laws enacted to protect the bird in certain countries are eminently sensible and thoroughly justified.

The bold and resolute tawny owl

The fork of a blasted tree is a well sheltered site frequently chosen by a pair of tawny owls to incubate their eggs and rear their young. The large, glittering coal-black eyes of the chicks, which look like little balls of white wool, gaze unflinchingly at intruders who try to reach the nest. The mantle of down is agreeably soft and warm to the touch and the fledglings are the very picture of delightful innocence. But those bright eyes already hint at qualities which will later become manifest, for the tawny owl (*Strix aluco*) is the most determined and courageous of all forest birds.

About the same size as the barn owl but giving the impression of being much more sturdy and powerful, the adult tawny owl, particularly the female defending her brood, is a brave and formidable adversary. It is indeed dangerous for anyone to try to approach the nest while she is raising her young. The famous British bird photographer Eric Hosking discovered this to his cost when he tried to take a picture of a female tawny owl and her chicks, provoking a furious attack that resulted in his losing the sight of one eye.

The tawny owl is a characteristic species of the deciduous forest but may also breed in Mediterranean-type or mixed woodland. The nest is generally situated in the hollow of a tree trunk, usually some distance from the ground, underneath a tangle of gnarled roots or in the side of an embankment. Less common nesting sites may include a hole in a wall, a rock cavity or even the inside of a chimney stack.

Given modern deforestation methods whereby venerable trees are systematically chopped down to make room for more rapidly growing younger species, tawny owls tend to make use of an old nest again and again, perhaps for ten consecutive years.

The female usually lays two to four eggs early in April, and

During the day the tawny owl perches on a sheltered branch, eyes half-closed, head tucked between the shoulders. It is difficult to see the bird against the bark of the tree trunk.

incubation, commencing as soon as the first of them is laid, lasts about a month. Number and timing, however, tend to vary from one region to another. In Sweden, for example, the tawny oil lays her eggs in December. When lemmings are abundant up to nine eggs may be laid for here, as in other areas, there is a strict inter-dependence between raptors and their rodent prey.

The tawny owls of Wytham Wood

A few miles to the north-east of Oxford lies Wytham Wood, scene of a survey by the ornithologist H. N. Southern of the territorial behaviour of a group of tawny owls. It also dealt with the influence of vegetational cover on the local small mammal population and considered the repercussions of such factors on the owls' reproductive habits.

In conjunction with his colleague V. Lowe, Southern marked on an ordnance survey map the precise area where each evening the characteristic hooting of the owls could be heard, some fifteen or twenty minutes before they set off on their hunting excursions. A few days later, having located the trees where the owls habitually perched at this hour, the naturalists forced them to fly away, noting that the birds never strayed beyond the bounds of a fairly restricted territory which coincided exactly with the zone previously marked on the map. To complete their investigations they caught several fieldmice and bank voles, ringing one paw of each animal before letting them go. Later they examined pellet contents and found a number of these rings, further confirmation of their theory that each owl had made its kill well within the confines of its territory – it being known that such rodents have a relatively limited range.

Facing page: The hollows of tree trunks are ideal nesting sites for owls. The modern policy of reafforestation often entails the planting of young trees with smooth, straight trunks that offer the birds no footholds and force them to leave the region altogether. It is the responsibility of forest authorities to provide roosts in such areas so that these valuable rodent-destroying raptors can continue to breed.

The extent of each owl's domain varied from about 20–30 acres but not all of them offered the same food possibilities. In areas where the undergrowth was fairly sparse fieldmice tended to be abundant and these formed the owls' principal prey; but in patches where there was more vegetational ground cover voles were the predominant species and the owls' diet would be correspondingly diversified. In some places, however, the bushes were so thick and closely packed that the birds were clearly unable to hunt at all despite the presence of quantities of rodents.

In the course of the Wytham Wood survey conditions altered radically from season to season. In the spring and summer of 1954, for example, hunting was good; but in the autumn of that year a serious myxomatosis outbreak decimated the rabbit population and all the local predators switched their attention to voles and fieldmice, drastically reducing the numbers of these rodents. This in turn caused feeding difficulties for most of the tawny owls in the region. Out of the eleven pairs studied by the naturalists only two – notably those living in an area where the food situation was not especially bad – were able to rear their young to maturity. Later on, however, as the amount of prey increased, seven of these pairs succeeded in safeguarding and preserving their broods.

Geographical distribution of the barn owl.

The experience and skill of the male is an important factor during the period of incubation for if he fails to attend to the needs of the female at this critical time she will be forced to interrupt her sitting and as a result the eggs may grow cold and deteriorate. Should they hatch successfully the male still has the added responsibility of providing food for the fledglings and if he is unable to do so they will die of hunger and perhaps be eaten by the parents. This is in fact what happened to the four remaining pairs of tawny owls under observation.

For the first two weeks after the owlets are born the mother remains with them in the nest. Later she leaves them to go out hunting but still keeps a watchful eye on their activities, ready to spring to their defence should they be menaced by intruders. Unable to fly until they are a month old, the owlets cling to the edges of the nest and gradually venture out onto nearby branches, still carefully supervised by the mother. The youngsters stay together for another two or three weeks, often perhaps on the same branch, but disperse in the evening to explore every corner of their parents' hunting grounds.

As the summer passes the family ties gradually slacken. Around the end of August, or even sooner if food is in short supply, the young tawny owls assert their independence and leave the nest. In their attempts to find territories of their own they may travel from five to thirty miles and some individuals have been known to journey up to two hundred miles before finally settling down. At the time of parting company with their parents the immature birds are not yet capable of catching rodents and therefore feed mainly on insects. But by the end of autumn, when insects are scarce, any owl which has not yet learned to hunt larger prey has little chance of surviving the cold season to come. Each winter between 47 and 67 per cent of young tawny owls fail to procure the necessary daily ration of 2–6 ounces of meat and die

BARN OWL
(Tyto alba)

Class: Aves
Order: Strigiformes
Family: Tytonidae
Total length: $13\frac{1}{2}$ inches (34 cm)
Wingspan: 36–$37\frac{1}{2}$ inches (91–95 cm)
Weight: $\frac{3}{4}$ lb (350 g)
Diet: flesh, mainly rodents
Number of eggs: 4–6
Incubation: 30–34 days

Adults
White, heart-shaped facial disc, black eyes. Back golden-buff with brown and grey streaks. Underparts white or yellowish with brown specks on breast and flanks.

Young
Down gives way to permanent plumage without any intermediate phase.

Food preferences of the tawny owl.

of starvation. The survivors, whether by skill or good fortune, become mature the following spring and when the forest is once more mantled in green their territorial hooting can be heard mingling with that of the adults at sunset.

Victims of the tawny owl

Dr Uttendörfer, the ornithologist who studied the behaviour of the eagle owl in the Carpathians, has also made a detailed survey of the food habits of the tawny owl in central Europe, using the same method of analysing the birds' pellets. Out of a total of 59,804 vertebrate traces he counted 39,457 mice and rats, 2,583 shrews, 979 voles, 400 young hares and rabbits, 129 bats, 32 weasels, twelve squirrels and eight stoats. Of the 8,452 bird specimens he calculated that the majority consisted of sparrows, followed by greenfinches, chaffinches, starlings, great tits, blackbirds, yellowhammers, etc. The remainder comprised 7,306 frogs and toads, 404 fishes and 42 reptiles. Invertebrates such as earthworms, slugs, snails, cockchafers and other beetles, mole-crickets and grasshoppers were also numerous.

In his English survey Southern counted 9,497 vertebrates, including 7,056 rodents, 1,358 shrews, 436 moles, 165 rabbits, four bats, four weasels and 474 birds (113 of them chaffinches). He also pointed out that in spring and winter prey consisted principally of rodents whereas in summer and autumn both smaller and larger animals predominated.

Other investigations of this nature carried out in different parts of Europe have confirmed that rodents make up 60–75 per cent, birds 5–20 per cent and frogs 3–12 per cent of all food eaten by the tawny owl throughout the year. It is evident, therefore, that the bird is a valuable ally to man, who can best ensure

The tawny owl plays a vitally important role in regulating the animal populations of woods and forests and maintaining an ecological balance for it is a specialised hunter which takes advantage of any species that happen to be prolific at any time.

The night-hunting owls of the world are to be found in many habitats and prey on a very wide range of animals, especially rodents. The following are all European species. 1. Snowy owl (*Nyctea scandiaca*). 2. Great grey owl (*Strix nebulosa*). 3. Hawk owl (*Surnia ulula*). 4. Tengmalm's owl (*Aegolius funereus*). 5. Ural owl (*Strix uralensis*). 6. Pygmy owl (*Glaucidium passerinum*). 7. Little owl (*Athene noctua*). 8. Tawny owl (*Strix aluco*). 9. Long-eared owl (*Asio otus*). 10. Scops owl (*Otus scops*). 11. Barn owl (*Tyto alba*). 12. Eagle owl (*Bubo bubo*). 13. Short-eared owl (*Asio flammeus*).

Facing page: The barn owl is so specialised a hunter of rodents that when the latter are particularly abundant the birth rate of the local owl population tends to increase.

its active cooperation by refraining from destroying its eggs, by sparing the fledglings and by calling off his daytime hunts for adult birds.

The endless hunt for vermin

Everyone knows that rats and mice are vermin but few people appreciate the enormous scope of the damage they can cause. The American scientist Ivan Sanderson has estimated that in the years following the Second World War rodents destroyed field crops equivalent in weight to all provisions sent abroad under relief schemes. The value of food in store which was destroyed by rats and mice also ran into hundreds of millions of dollars. In addition rodents did incalculable harm in other spheres such as public utilities (destroying the casing of gas pipes and electricity cables), and damaged many precious works of art.

The owls of the world, which have adapted to diverse habitats, are by far the most efficient hunters of vermin. The barn owl extends its activities to barns and buildings in villages and town suburbs. The tawny owl and Scops owl keep watch in orchards, groves and fields. The eagle owl, probably the most skilful hunter of the lot, operates over any kind of open ground; and the short-eared owl is an inhabitant of swamps and marshes.

It is essential, in order to increase agricultural productivity, to keep vermin in check and anyone who works on the land should recognise that an owl is a friend and not an enemy. The more scientific knowledge we can obtain about its habits the quicker we can dispel the mists of superstition that have for too long shrouded the real facts.

ORDER: Strigiformes

The owls belonging to the order Strigiformes vary considerably in size and weight, ranging from the sturdy eagle owl (*Bubo bubo*) which may measure up to 28 inches, to the pygmy owl (*Glaucidium passerinum*) which is only about 6 inches long.

Insofar as it is possible to generalise, owls are distinguished by a large head (sometimes adorned with ear-tufts), a short neck, a strong and compact body and a fairly short tail. The plumage is normally rather drab, in which brown, grey and ochre tend to be predominant. A notable exception is the snowy owl which is mainly white, but with variable dusty brown markings that are darker and more conspicuous in the female.

The eyes of a typical owl are large and frontally placed so that the bird possesses a fairly wide field of binocular vision in the direction it is facing but a very limited range of lateral vision. The retina is well provided with rods–cells that are sensitive to light intensity–so that the bird can see perfectly in the half-light of dawn or dusk. It can gauge distances and contours quite well but has difficulty in focusing on objects close at hand. The head is remarkably flexible and can be revolved through three-quarters of a circle (270°), compensating for any other visual deficiencies. The eyes are surrounded by large facial discs made up of thin radiating feathers. In the barn owl these run together to make the shape of a heart.

Hearing is extraordinarily well developed so that an owl can pick up a wide range of low and high-pitched sounds. The auricular slits, protected by folds of skin that can be opened or closed, may be very large and are not symmetrically placed so that the bird can easily locate the exact spot from which a sound emanates.

The feet are of moderate size and are furnished with four toes, each with strong, curved and pointed claws. The outer toe is very mobile and can be turned either towards the front or the rear.

Most of the species comprising this order are active either at dusk or at night; some of them, however, such as the hawk owl (*Surnia ulula*) and the snowy owl (*Nyctea scandiaca*) also hunt during the day. Almost all are sedentary but the Scops owl (*Otus scops*), the long-eared owl (*Asio otus*) and the short-eared owl (*Asio flammeus*), as well as a few other species, tend to move southwards in the winter.

Depending on individual size and weight, owls hunt an extremely varied range of prey, from insects to hares and rabbits. The largest part of the diet, however, usually consists of rodents. The birds' controlling influence on vermin in rural areas is of vital importance. The abundance or scarcity of rats, mice and other rodents in a particular district has a bearing on the birds reproductive behaviour. When these small mammals are plentiful the female owls lay a greater number of eggs and a larger proportion of young birds manage to survive the winter.

Some owls have specialised hunting tastes. The fish owls of genus *Ketupa*, for example, feed principally on fishes and crustaceans.

The Strigiformes have a vast global distribution and are divided into two families, Tytonidae and Strigidae.

The Tytonidae comprise the subfamilies Tytoninae and Phodilinae. In the former are some ten species, including the barn owl (*Tyto alba*). The two species making up the latter subfamily are the bay owl (*Phodilus badius*) and the Tanzanian bay owl (*Phodilus prigoginei*).

The Strigidae comprise about twenty genera and are in turn divided into two subfamilies. The Buboninae include the eagle owl, Scops owl, fish owl and little owl (*Athene noctua*). Among the Striginae are the tawny owl (*Strix aluco*), short-eared owl, long-eared owl, etc.

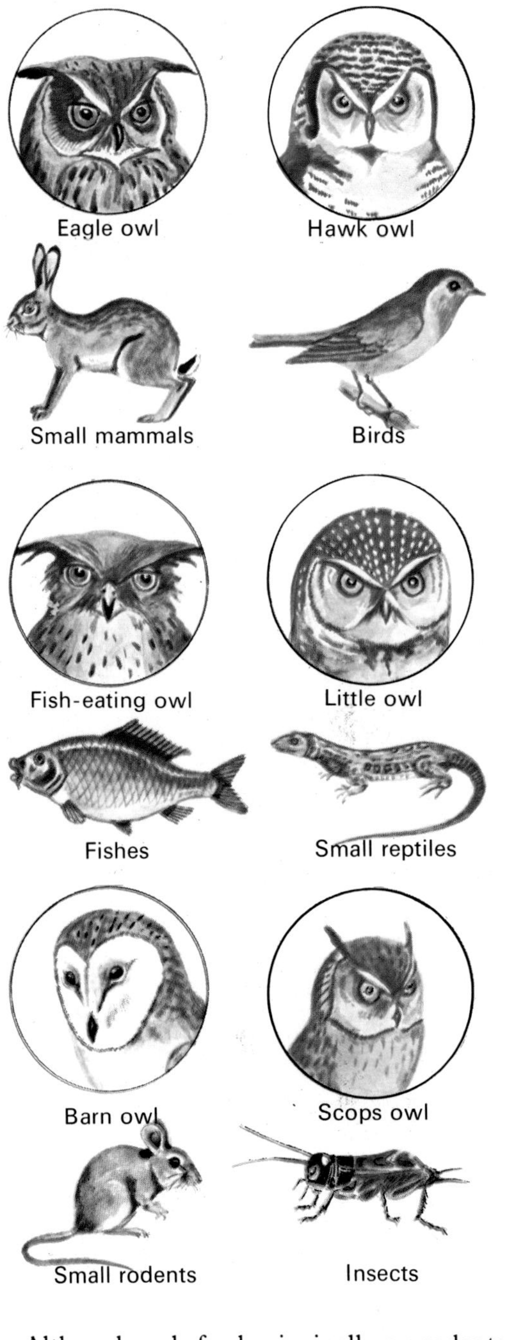

Although owls feed principally on rodents, other small mammals, birds and reptiles also figure prominently in their diet. The eagle owl consumes large quantities of insects and the fish-eating owl of southern Asia, as its name suggests, hunts fishes and crustaceans.

Facing page: The Strigiformes are found all over the world apart from some islands in the Pacific. The order is divided into two families, the Tytonidae and the Strigidae. The short-eared owl (above) and the snowy owl (below) belong to the Strigidae.

CHAPTER 13

Birds of the deciduous forest

In spring the forest trees are once again decked in fresh green leaves and the air is filled with the noisy chattering of myriads of birds. The traditional woodland songsters–including nightingales, warblers, blackbirds, fieldfares and orioles–trill and twitter among the foliage, but not for the sheer love of vocal exercise. Like many other animals the bird inhabitants of the deciduous forest are conscious of their territorial rights and the characteristic calls that make up their 'song' are the means by which the males of the various species proclaim individual ownership and issue warning to rivals that they are if necessary prepared to defend their frontiers.

All so-called territorial animals announce their rights and intentions in typical ways. Mammals, endowed with a highly developed sense of smell, often use olfactory signals–streams of urine, heaps of excrement and glandular secretions which can be sprayed on bark, leaves and grass. Other animals use visual methods (conspicuous colours and patterns) to signify their presence to adversaries; notable examples are the marine fishes of tropical coral reefs whose gleaming multicoloured scales are clear identifying signals to others of their kind.

The birds of wood and forest, however, have virtually no sense of smell; furthermore the thick curtain of leaves and tangled undergrowth make it impossible to see one another at a distance so that the only way for a male to establish his claim to territory is to proclaim it vocally. The amazing tonal and rhythmical variety of sounds given out by the innumerable woodland bird species are the equivalents of the striking colours and markings or the distinctive scents of other animals which are normally active in more exposed habitats.

Amid the babel of warbling and chirping of so many forest

Facing page: The nuthatch, like the tree-creeper and woodpecker, seeks its insect prey on tree trunks but does not need to use the tail for its acrobatic movements, gripping the surface with its feet, which have short tarsi and particularly strong claws. Ringing has shown that it is a sedentary species.

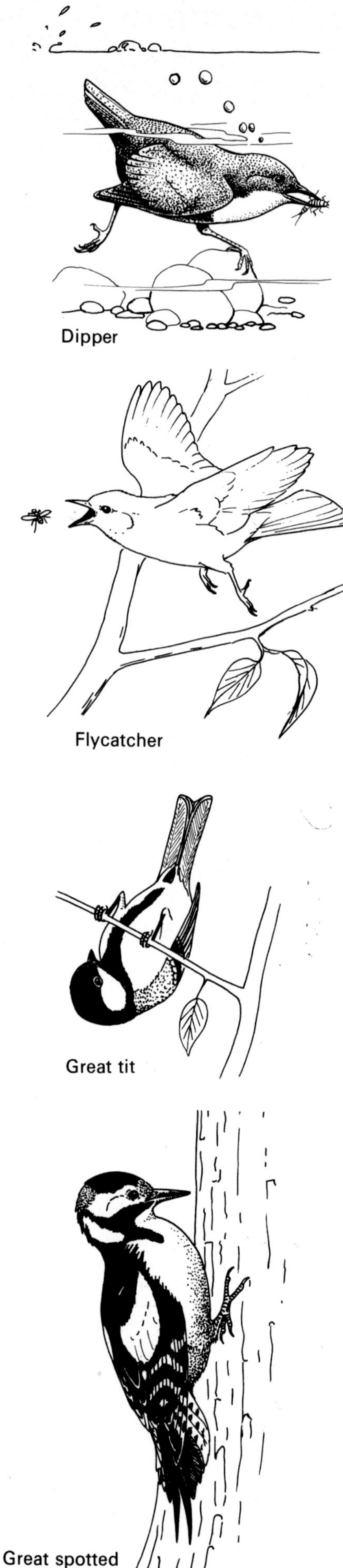

birds, each male has to get his message across to competitors in his own characteristic way. The unmistakable warning call of the male nightingale from the thickets automatically deters would-be intruders of his own species; but he will be quite indifferent to the incessant chatter of warblers, wrens and blackbirds close by, for the territories of the different species frequently overlap. Rivalry is strictly confined to members of the same species, the objective being to gain possession of a piece of terrain in which all available food can be monopolised. Thus when a blackbird emerges victorious from a battle with a neighbour in the undergrowth he will have won command of a private hunting preserve denied to all other male blackbirds. Having solved the food problem he can then turn his attention to other urgent matters such as attracting a female, making a nest and rearing a family,

Young blue tits, like the young of other tit species, do not linger long in the nest but leave it when they are about three weeks old. The adults continue to feed them for a while but soon they are independent enough to fly off and fend for themselves.

further valid reasons for his refusal to tolerate the intrusion of potential rivals within his territory; and when he is not occupied with hunting for food he again resorts to song in order to avoid confrontations with other blackbirds. The reason why this jealous male will put up with the presence of a nightingale or woodcock nearby is that these birds pose no threat to his welfare. They occupy different ecological niches, not competing directly with him for food. Their tastes are not necessarily his and even when they set their sights on the same kind of prey they will use different hunting methods and be active at other levels of the forest.

Once he has achieved his territorial aims the breeding season is at hand and the vocal efforts of the male are directed towards attracting a female. A bare fifty years ago it was still popularly

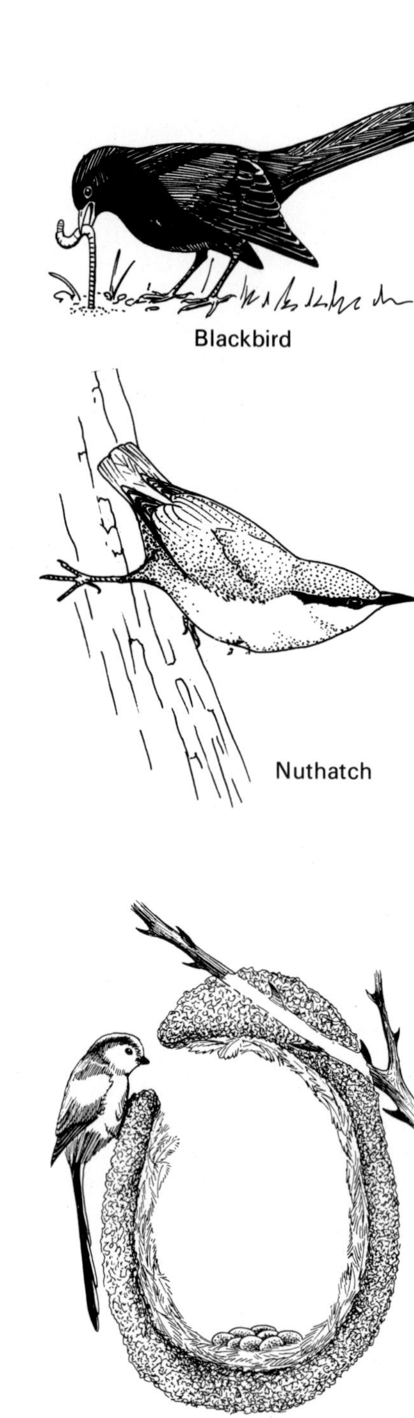

Blackbird

Nuthatch

Nest of long-tailed tit

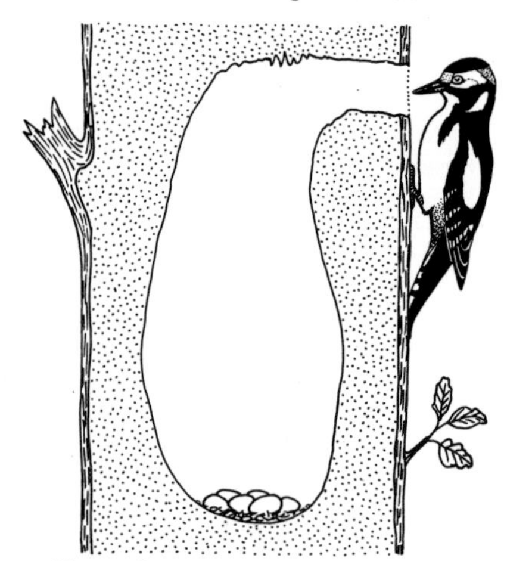

Nest of great spotted woodpecker

believed that these clamorous birds were joyously heralding the reappearance of spring. The group of British ornithologists who first revealed the scientific facts about the territorial behaviour of birds had no intention of invalidating the traditional romantic notions of the great poets of the past; and although zoologists have a duty to recognise and publicise such facts, poets of the future, it is to be hoped, will continue to express their feelings on the subject in the time-honoured terms of fancy and imagination.

Some forest birds, such as robins and orioles, have the best of both worlds in that they can stake out territory in two ways, by voice and appearance. In addition to being indefatigable singers they also have brightly coloured plumage which immediately distinguishes them from other species. The red breast of the robin, especially when caught by the sun, stands out conspicuously against the background of leaves as a clear visual signal. Experiments have shown that one only has to drop a bunch of red feathers on a robin's territory to unleash a furious assault by the dominant male concerned.

Such stratagems, whether vocal or visual, may well discourage rivals but they do not necessarily provide a safeguard against enemies–quite the reverse. Predators are skilled in detecting the faintest sounds and the bird which shatters the morning calm by trilling to his mate or puffing out his gaudy breast feathers runs the risk of betraying his presence to an animal bent on catching and killing him. The interesting point here is that tragedy for an individual may spell hope for the community. The forest hunters are instinctively drawn towards the ostentatious males and usually ignore the drabber, quieter females. Moreover, once they have fulfilled their reproductive responsibilities, the males perform a further service by luring predators away from mothers and fledglings, even sacrificing their lives and thereby safeguarding the new generation.

Hunting grounds for all

The deciduous forest is a paradise for a multitude of insectivorous and omnivorous birds. Within an area of say one square foot down to a depth of only one inch there may be up to 1,500 living creatures, not counting the teeming population of micro-organisms to be found in any random handful of earth. In addition to the invertebrates feeding on dead leaves and humus there are those that live in tree trunks, on branches and leaves, flowers and fruit. The forest floor is the realm of earthworms, snails, spiders and countless insects, some useful, others harmful, but all constituting food for a variety of birds which concentrate on hunting them in spring and summer but, if they do not migrate, change in winter to a mixed diet of larvae and seeds.

How is it that in such a complex bird community of insectivores and omnivores the natural equilibrium of the environment is maintained? How can the danger be averted of so many winged hunters exhausting the forest's food supply in the spring–something that would jeopardise the future of fledglings still too small to procure their own food or to fly off to other regions?

The golden oriole is one of the most attractive and characteristic of European woodland birds, though seldom seen in Britain. The nest, built by both male and female, is usually suspended from the fork of a high branch, away from the trunk.

The birds themselves have evolved solutions to these vital problems. Each species avoids rivalry with another by dividing the forest space into territories, the extent of these depending on the abundance of prey. Thus the domain of a nightingale will be more restricted in a good season when there are plenty of insects than in a time of drought when they are scarce. But not only do these bird species, like larger predators, keep up steady pressure on the insect community but each one finds victims at a separate level. Some birds peck away at the ground, some rake underneath dead leaves, others drill into bark. They will thus uncover all the hiding places of their tiny victims without ever interfering with one another's activities. This is their secret of survival–a multiplicity of hunting grounds, sometimes shared by several species but exclusive to one individual within that species, and the fullest exploitation of different levels of the forest by its inhabitants.

Birds of the forest floor

In the age-old battle between birds and insects many of the latter have sought refuge in unaccustomed places as, for example, in water. Thus certain beetles conceal and camouflage their eggs

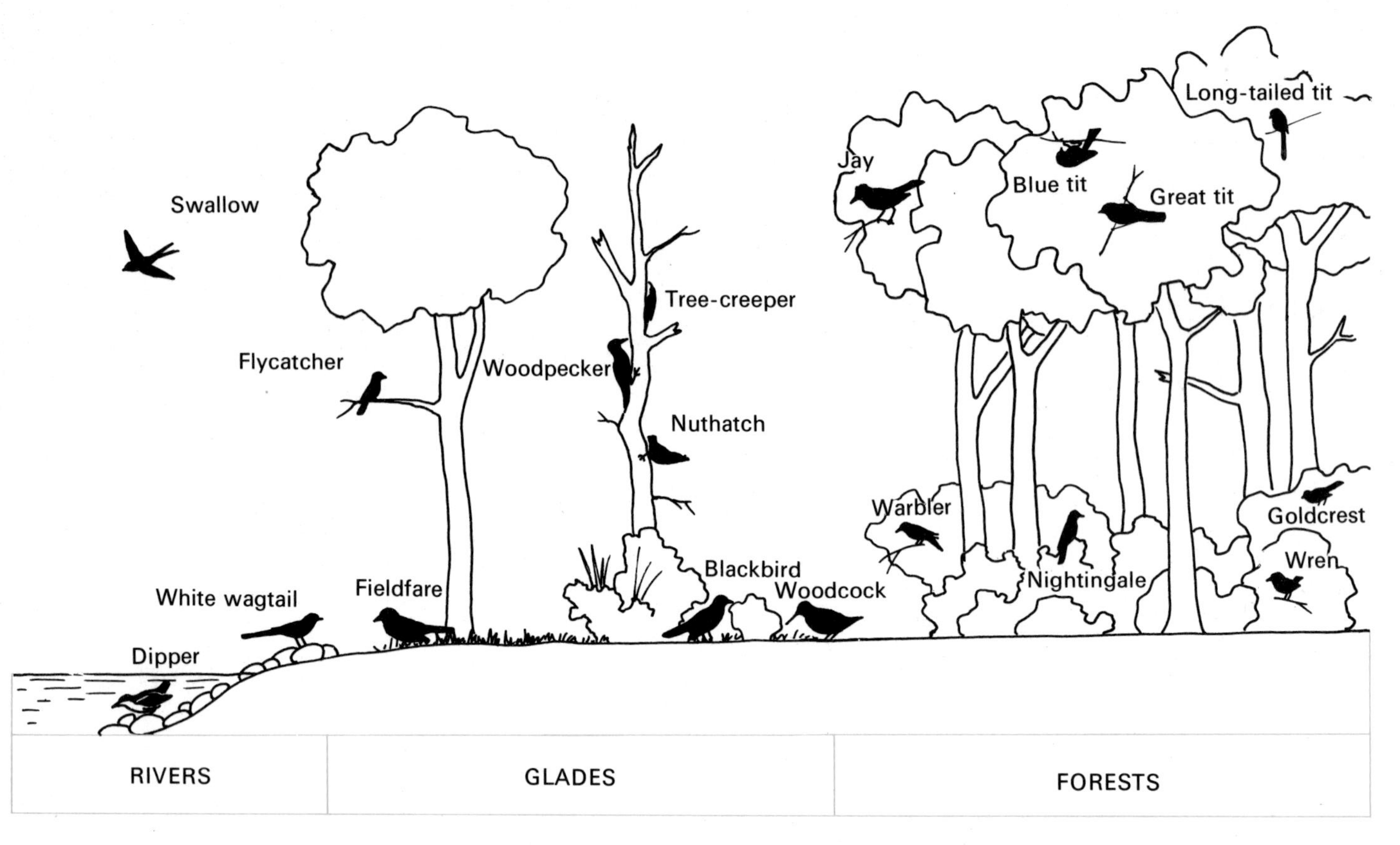

The birds of the forest take the fullest advantage of their varied surroundings and even when they feed on similar substances operate at different levels so that they do not compete directly with one another. This diagram shows how different species hunt for food in mid-air, in water, on bare bark, among the leaves of tall trees, in woodland clearings and in thick undergrowth.

Facing page (above) : Warblers are tree-dwelling birds which seldom venture far from cover in quest of food. (*Below*) The blackbird has a much more varied food range and frequently leaves the forest to scour for prey on the ground, in woodland glades and even on garden lawns.

in the pebbly bottom of streams and rivers. But this ruse is only partially successful for even here the beetles have a formidable enemy in the dipper, an agile bird which has adapted to these surroundings and fears no competitors as it dives repeatedly into the fast-flowing torrents to fish for arthropods.

A frequent visitor to a woodland clearing traversed by a river is the fieldfare, a member of the thrush family. It is thus related to the blackbird but the two birds are not rivals for the fieldfare hunts earthworms and other invertebrates on open grassy ground whereas the blackbird is normally more active in undergrowth, on river banks and in ditches. Both birds have a partiality for worms of the class Annelida and for insect larvae on the ground or under a shallow layer of leaves, raking the latter with their feet. This type of food is also eaten by the woodcock but here again there is little competition for this bird drills several inches below the surface with its long bill—an area unattainable to the fieldfare and blackbird.

The hunting grounds of these three species which feed so voraciously on soft-bodied, medium-sized invertebrates are delimited by zones of thick bush and scrub where, because of their relatively large size, easy movement is impeded. Such areas are, however, the province of the nightingale and various warblers, smaller birds which can thrust their way into the underbrush to dislodge those insects which constitute their specialised diet. But there are also limits to the activities of these nimble birds, notably dense thickets of thorn and bramble where there would not appear to be room even for a snake to glide. Nevertheless, there is one bird, tiniest of all the winged residents of the forest, which is not deterred in the slightest by thorn bushes, namely the wren. This lively little bird rules a kingdom of its own, flitting through the remotest parts of the undergrowth in search of insects and spiders which no other bird can possibly catch.

Specialists and opportunists

Just as the birds on the 'ground floor' of the forest go about their hunting activities without bothering one another, so too do the residents of the 'upper storeys'–the birds that live, feed and raise their young in the trees. The tits, for example, are the acrobats of the forest, flitting through the foliage, inspecting each leaf, often hanging head-downward from a thin branch, probing and tapping with their beaks until they discover a tasty chrysalis. No small animal, however cleverly camouflaged, can long escape their attentions. Flycatchers too will often be seen alongside tits on the edge of woodland clearings but do not challenge the latter for food for they are quite uninterested in insects crawling about on tree trunks or leaves. The agile flycatchers hunt much in the same way as swallows, using a bare branch as a watch post, waiting patiently until they spy a gnat or a butterfly some distance away, then chasing the insect and capturing it in mid-air.

Deep in the forest there are a number of possible hiding places for insects–soft earth, the intersection of a couple of branches, a hollow tree trunk or a fissure in rough bark where they sometimes bore long tunnels. Yet no insect is completely safe for there always seems to be a predator lurking nearby, specialising in the necessary hunting technique. Insects that live on the bare trunk of a tree are usually well camouflaged but still fail to escape the attentions of the nuthatch, a bird that can walk up and down a perpendicular trunk with consummate ease and agility, thanks to the unusual structure of its legs and the

The wren, the spotted flycatcher, the nightingale (*left to right*) and the blackcap (*below*) do not compete for food for each hunts in different places and in different ways. The fledglings of all these species react to their parents' arrival at the nest by opening their mouths wide. Experiments have been made to find out what stimulates this action. In very early life it appears, in a number of species, to be a matter of touch, for the effect can be achieved by lightly shaking the nest. Later a visual stimulus is needed, such as the actual presence of the adult bird or its close replica.

BLACKBIRD
(Turdus merula)

Class: Aves
Order: Passeriformes
Family: Muscicapidae
Total length: $9\frac{1}{2}$–10 inches (24–25 cm)
Wing-length: $4\frac{1}{2}$–$5\frac{1}{2}$ inches (11·7–13·8 cm)
Weight: 3–4 ounces (75–120 g)
Diet: omnivorous
Number of eggs: 3–5, sometimes 6

Male completely black except for yellow-orange ring round eye and yellow-orange bill. Female dark brown above, red-brown with darker markings on breast; brown bill.

ROBIN
(Erithacus rubecula)

Class: Aves
Order: Passeriformes
Family: Muscicapidae
Total length: $4\frac{3}{4}$–$5\frac{1}{2}$ inches (12–14 cm)
Wing-length: $2\frac{1}{2}$–3 inches (6·7–7·6 cm)
Weight: $\frac{1}{2}$–$\frac{3}{4}$ ounce (13–19 g)
Diet: omnivorous
Number of eggs: 5–7
Incubation: 12–15 days

Breast, forehead and sides of head bright red. Back olive-brown, belly white.

WREN
(Troglodytes troglodytes)

Class: Aves
Order: Passeriformes
Family: Troglodytidae
Total length: $3\frac{1}{2}$–4 inches (9–10 cm)
Wing-length: $1\frac{3}{4}$–2 inches (4·3–5 cm)
Weight: $\frac{1}{4}$–$\frac{1}{3}$ ounce (8–11 g)
Diet: omnivorous
Number of eggs: 5–6
Incubation: 14–16 days

Brown plumage, darker on back and barred on wings and tail. The tail is always held vertically.

NIGHTINGALE
(Luscinia megarhynchos)

Class: Aves
Order: Passeriformes
Family: Muscicapidae
Total length: $6\frac{1}{2}$ inches (16–17 cm)
Wing-length: 3–$3\frac{1}{2}$ inches (7·8–9 cm)
Weight: $\frac{3}{4}$–1 ounce (20–25 g)
Diet: omnivorous
Number of eggs: 4–5
Incubation: 13–14 days

Upper parts uniformly dark brown, underparts whitish; fairly long tail rufous.

gripping power of its claws. Insects that seek refuge in cracks of the bark are efficiently scooped out by the tree-creeper with the aid of its long, slender, curved bill. The xylophagous (wood-eating) insects that conceal their larvae even deeper in the trunk of the tree are at the mercy of the black woodpecker whose beak has the strength and sharpness of a tiny scalpel, ripping away the top layer of wood to uncover the entry to tunnels bored by its insect victims. The intruder then extends its long, probing tongue (the tip of which is horny and barbed) and feeds at leisure. Other woodpecker species search for food in a similar way, pecking at the tree trunks at different heights.

These specialised hunters who find the deciduous forest such a rich source of sustenance are often seen in the company of other insectivorous and omnivorous birds which are more or less permanent woodland residents. Goldcrests follow the great tits and wrens; wagtails are busy in fields and plains or on the banks of streams alongside blackbirds and fieldfares; stonechats and whinchats mingle with other characteristic scrub species; shrikes impale their victims on thorn bushes; and even buntings and sparrows will venture to the fringes of a wood in springtime to look for invertebrates.

We have taken a brief glance at some of the most characteristic forest birds and noted the different ways in which they go about finding their food as well as the different levels at which they operate. As has already been implied, there is a further broad distinction which is convenient for placing them in one of two categories, depending on whether they can be classified as specialists or opportunists in their hunting activities.

In the first group are those birds that are well adapted to exploit a particular environment–the dipper, for example, which finds all its food in water, or the woodpecker which drills deep into tree trunks. Their spheres of activity are comparatively narrow and restricted and they would be incapable of adapting to other conditions outside the natural surroundings where they have evolved.

The birds of the second group have more eclectic tastes and will feed on almost anything that happens to be available. Their hunting grounds may overlap with those of certain specialists. The most typical opportunist is without doubt the jay, a member of the crow family, which will pursue its prey through the foliage or pounce at it on the ground or on a dry tree stump. This resourceful and attractively coloured bird will also feed on vegetable matter, acorns making up about half of its diet.

Some birds are freed from the rivalry of other species by reason of their digestive capacities. Thus the cuckoo, whose monotonous call echoes through the woods in spring and which keeps company with other insect-eating birds, specialises in catching processionary caterpillars which are notoriously difficult to swallow because of their stiff, spiny protective covering. The cuckoo's secret is a stomach which can compress and periodically regurgitate the caterpillars' indigestible bristles.

The more naturalists are able to find out concerning the habits of these delightful forest birds the more influence they will be able to exert on those who still continue to show lack of wisdom and consideration in their meddlings with the natural scene. Enough has been said already about pesticides, which pose the greatest threat to birds of field and forest. The possible repercussions of these poisonous substances are incalculable but there are signs that the dangers are now being recognised so that effective measures can be taken to avert catastrophe. Yet there are other, less obvious threats which can only be dealt with by the authorities responsible for reafforestation programmes in different countries.

In the course of millions of years the deciduous forest biome has evolved as a complete biological entity in which predators and prey are perfectly in balance. The delicate equilibrium could only be upset by the occasional natural catastrophe until man accelerated the process by sacrificing the forest to his economic needs, with disastrous consequences in many areas. After much trial and error experts in silviculture have come to realise that by replanting a forest with a single tree species–usually a conifer–many insect-eating birds are automatically deprived

The robin (*left*) prefers to build a nest in the shade of undergrowth, often on a bed of dead leaves. The tree-creeper (*right*) which uses its slender curved bill to explore cracks in the tree trunk for food, can similarly find suitable cavities in the bark for nesting.

TITS

Class: Aves
Order: Passeriformes
Family: Paridae
Diet: basically insects
Number of eggs: 8–14
Incubation: 12–15 days

GREAT TIT
(Parus major)

Total length: $5\frac{1}{2}$ inches (14 cm)
Wing-length: $2\frac{3}{4}$–$3\frac{1}{4}$ inches (7–8·2 cm)
Weight: $\frac{1}{2}$–$\frac{3}{4}$ ounce (15–22 g)

Adults
Head and neck black, cheeks white. Upper parts grey-green, breast and belly yellow with black central stripe.

Young
Yellowish cheeks, top of head brownish-black.

BLUE TIT
(Parus caeruleus)

Total length: $4\frac{1}{4}$–$4\frac{3}{4}$ inches (11–12 cm)
Wing-length: $2\frac{1}{4}$–$2\frac{3}{4}$ inches (6–7·2 cm)
Weight: $\frac{1}{3}$–$\frac{1}{2}$ ounce (9–15 g)

Adults
Top of head, wings and tail blue. Back greenish, breast and belly yellow. Cheeks white. A black stripe extends from either side of the bill across the eyes and down the neck, the lines meeting around nape and cheek to blue-black chin.

Young
Sides of the head yellow.

LONG-TAILED TIT
(Aegithalos caudatus)

Total length: $5\frac{1}{2}$ inches (14 cm)
Wing-length: $2\frac{1}{4}$–$2\frac{3}{4}$ inches (5·5–7 cm)
Weight: $\frac{1}{4}$–$\frac{1}{3}$ ounce (6–10 g)

Adults
Well contrasted plumage. Head white with black stripe above each eye. Back almost black with pink areas. Underparts white, breast and belly pinkish. Long tail is black and white, comprising feathers of uneven length. There are a number of subspecies which vary considerably in pattern and colour of plumage.

Young
Sides of head and back brown.

Facing page: The long-tailed tit (*above, left*) is noted for its nest-building skill. The great tit (*above, right*) and the blue tit (*below*) are intelligent and extremely acrobatic species which are often seen in parks and gardens, using highly ingenious methods of finding food.

of sites where they have traditionally nested and found food. Woodland birds are unable to breed in districts planted with straight, smooth-barked trees and it is important to make provision for them by establishing artificial roosts in such areas. One way or another, it is in our interest to preserve these forest species which by feeding on noxious insects keep the environment clean and healthy and by so doing stimulate new vegetational growth.

Woodland acrobats

The tits of the genus *Parus* are nimble, restless birds, continuously on the move. Their diet consists basically of insects and their active metabolism hardly allows them to be still for a moment. In the course of their incessant search for insects and larvae they adopt the most unlikely attitudes, hanging downwards from branches as they explore every leaf with rapid, nervous pecks. In autumn they switch their attention to seeds and grains, especially those with a rich fat content.

In Europe the commonest member of the family–and the largest too–is the great tit (*Parus major*), a familiar resident of parks and gardens, even in winter, sleeping in fissures and holes. The pattern of life changes in spring as each male establishes his territory, attracting the female with food offerings, to which she responds with infantile chirps and wing fluttering. In March they build a nest in a tree hollow, piling up blades of grass, twigs, roots and other materials, and often adding scraps of wool or hair. The first egg will be laid before the nest is completed and up to ten, twelve or even fourteen are then laid at two-day intervals. There may later be a second brood, but in the course of the year many fledglings die–sometimes as many as 87 per cent. The death rate among adults is also high and few tits live for more than about nine years.

The male feeds his companion during the incubation period and the young hatch after two weeks. At first they are given soft foods such as butterflies' eggs but they gradually progress to tougher, more substantial items. Within 18–20 days, sometimes sooner, they are flying around with their parents and demanding food incessantly by chirping and flapping their wings.

Family life continues for two weeks more and then the youngsters go their own ways to find their own territory. This may be fairly close to their birthplace but is frequently a good distance away–in fact, anything up to a hundred miles.

Tits are clever birds with a great sense of curiosity. In Britain, where it is the custom for bottles of milk to be delivered to houses and left outside on the doorsteps, it is not uncommon for tits to peck through and even remove the foil caps of these bottles to enjoy a refreshing drink. The success of the first enterprising bird is enough to encourage its companions to emulate its example. So entertaining are these birds that it is almost a national custom in Britain for people to hang a coconut, a bag of nuts or some scraps of fat from the bird-table so that great tits and blue tits can find a regular supply of food during the winter.

Facing page : After spending several days concealed in the burrow drilled by its parents the young green woodpecker soon learns to poke its head out of the hole and noisily demand food.

Miss Len Howard, author of a book entitled *Birds as Individuals*, lives in Sussex, in southern England. The birds of the surrounding woodlands are her friends and they have become accustomed to her presence. In the course of her observations she has remarked that the individuals of a single species do not necessarily behave in identical ways and that each has a recognisable personality. Her studies of the behaviour of great tits, for example, reveal that some birds pair for life while others, more fickle, continually change partners.

Another popular assumption which has been exploded is that song is the exclusive prerogative of the male. It is true that most females are mute but some are known to sing even more energetically and tunefully than their mates. According to studies by Terry Gompertz, published in 1961, great tits have a repertory of about eighteen distinct calls, from the feeble chirpings of the nestlings to the strident love calls and territorial proclamations of the adult birds.

The long-tailed tit (*Aegithalos caudatus*) is not strictly speaking a member of the tit family since it does not belong to genus *Parus*, but it goes by that name because of its many resemblances to the 'genuine' tits. Very gregarious, these birds form small groups which frequent forests and keep together by repeatedly calling to one another. The breeding season starts very early. From February onwards the little groups disperse and each couple settles in its own territory. It takes them three weeks to build a nest–an elaborate pear-shaped construction with a side entrance in the upper section. It is extremely large in relation to its occupants–measuring over seven inches high and four inches wide. The interior is provided with a layer of feathers which are meticulously carried in one by one. Ornithologists have estimated that the birds may make between one and two thousand journeys during their nest-building activities. The female eventually lays between seven and twelve eggs. The chicks leave the nest as soon as they can, flying to and fro in the manner of great tits, under the initial guidance of their parents.

In addition to the great tit and the long-tailed tit, other members of the family commonly found in Europe and Asia include the blue tit (*Parus caeruleus*), with blue wings and head; the coal tit (*Parus ater*) which likes to nest in old, worm-eaten trunks but will in fact make do with any kind of hole; the marsh tit (*Parus palustris*) which despite its name does not normally frequent marsh or swampland; the crested tit (*Parus cristatus*), an inhabitant of the coniferous forest; the penduline tit (*Remiz pendulinus*), a resident of reed-beds bordering swamps; and the bearded tit (*Panurus biarmicus*), whose place in systematics is rather uncertain and which is an inhabitant of wetlands as well as other districts.

In North America the tit is known correctly by its full name 'titmouse' or as the 'chickadee'. The black-capped chickadee (*Parus atricapillus*) is thus the same bird as the Eurasian willow tit; but exclusive inhabitants of the New World include the Carolina chickadee (*Parus carolinensis*), the brown-capped chickadee (*Parus hudsonicus*) and the tufted titmouse (*Parus bicolor*), closely resembling the crested tit in appearance.

WOODPECKERS

Class: Aves
Order: Piciformes
Family: Picidae
Diet: mainly insects
Number of eggs: 3–9
Incubation: 12–19 days

GREEN WOODPECKER
(*Picus viridis*)

Total length: $12–12\frac{3}{4}$ inches (30–32 cm)
Wing-length: $6\frac{1}{4}–6\frac{3}{4}$ inches (15·7–17 cm)
Weight: $5–7\frac{1}{2}$ ounces (150–210 g)

Adults
General colour green with grey tones on belly; rump yellowish-green. Crown of head red; facial mask black. Moustachial stripes black, bright red in centre. Short, stiff tail.

Young
Mottled plumage on back and breast. No facial mask; crown of head dull red; moustachial stripes barely visible.

PIED OR GREAT SPOTTED WOODPECKER
(*Dendrocopos major*)

Total length: $8\frac{1}{2}–9\frac{1}{4}$ inches (21–23 cm)
Wing-length: $5\frac{1}{4}–6$ inches (13–15 cm)
Weight: $2\frac{1}{2}–3\frac{1}{2}$ ounces (70–100 g)

Adults
Large white marks on shoulders. Cheeks white with black borders that extend over back and wings. Crown of head black; nape of male red. Lower part of belly and lower tail coverts bright red.

Young
Crown red; lower tail coverts a duller red than in adults.

BLACK WOODPECKER
(*Dryocopus martius*)

Total length: $18–18\frac{3}{4}$ inches (45–47 cm)
Wing-length: $9\frac{1}{4}–10\frac{1}{4}$ inches (23–26 cm)
Weight: 9–11 ounces (245–315 g)

Adults
Male is larger than any other European woodpecker. Plumage uniformly black except for red feathers forming a kind of crest on the crown. There is also a small red mark on the neck of the female.

Young
Plumage dark chestnut rather than black.

The tree-tapping woodpeckers

The woodpeckers of the family Picidae feed almost wholly on xylophagous insects. There are many species in the Holarctic region but probably the best known is the green woodpecker (*Picus viridis*). Holding its body upright, gripping the rough projections of bark with its claws (two of which are directed forwards and two to the rear) and supporting itself with a tail consisting of stiff feathers, the bird moves continuously up and down the tree trunk in a series of tiny hops. The tail feathers are moulted in an unusual manner for the two central rectrices do not drop until the outer tail feathers are sufficiently strong to provide the necessary support. As it climbs the woodpecker scours the surface of the trunk for insects, pausing periodically to hammer away at the bark with its sharp bill. Having uncovered an insect or larva in the hole which it has drilled, the bird promptly extends its long, slender, sticky tongue (pointed like a dart at the tip), limes its victim and immediately swallows it. It then continues its exploration and tries its luck elsewhere.

In addition to insects in and on the tree trunk, woodpeckers will also hunt ants on the ground while some birds even feed on plants or suck the sap of certain species. In common with other members of the family, the green woodpecker's flight pattern is characteristically undulating, with short sequences of four or five rapid wing beats alternating with long dipping glides. Continuously dipping and climbing, the woodpecker pursues a direct forward course but cannot keep it up over a long distance.

Woodpeckers are not sociable birds and generally lead a solitary life, intolerant of the intrusion of their congeners as they patrol vast territories. Even a breeding pair sharing the same hunting grounds tend to avoid each other's presence when not engaged in reproductive activity. To proclaim their territorial rights and warn off rivals they give out strident cries, especially during the spring, which can be heard over a wide area. Many woodpeckers also tap their bills rhythmically against tree trunks but the sound produced is a drumming unlike the characteristic hammering which they make when they are searching for food. Nor do any two species drum away at precisely the same pitch, rhythm or speed.

Although acoustic signals are more effective in thick forest and woodland, woodpeckers also communicate by visual means, most of them being vividly and conspicuously coloured in black and red, black, white and red, green and red, etc.

Woodpeckers make good use of their long pickaxe-like bills to chisel cavities in trees for their nests. One might assume that by so doing they would cause considerable damage to the trunks but in fact the birds, in response to the simple law of expending the least amount of energy to achieve a given end, habitually choose surfaces that are dry, parasitised by fungi or already pitted and scarred.

Nevertheless, the allegation that they cause damage cannot be completely discounted. In forests made up only of healthy trees they have no alternative but to drill into sound, hard wood. Conifers are seldom chosen for they are full of resin, so that these

trees are usually left intact while the trunks of neighbouring trees, such as poplars, are deeply incised.

The shape of the cavity selected as a nesting site obviously varies enormously but a typical nest will consist of a narrow, more or less circular tunnel, running horizontally or perhaps slightly upwards and terminating in a fairly deep, oval chamber, the lower end of which is carpeted with small slivers of wood. Here the female lays from three to nine shiny white eggs (the number of course varies according to species). The absence of colours and markings on the shells is because woodpeckers, like all birds that nest in holes, have no need for camouflaged eggs.

Incubation and rearing of the young are both accomplished very speedily. The nestlings are born naked and blind but their behaviour is totally contrasted to that of other families. Particularly sensitive spots on either side of the bill enable the young birds to feel the touch of the adults, signalling that food is about to be offered. At first the chicks remain squatting at the bottom

The pied or great spotted woodpecker, like all members of the family, has a strong, straight bill with which it brings food for the hungry offspring. Its main use, however, is to drill a nest in a tree trunk at breeding time. The bird's reputation for causing damage to bark is greatly exaggerated for it normally nests only in dry or diseased trunks.

Breeding and wintering zones of woodcock.

WOODCOCK
(Scolopax rusticola)

Class: Aves
Order: Charadriiformes
Family: Scolopacidae
Total length: $13\frac{1}{2}$ inches (34 cm)
Length of bill: $2\frac{1}{2}$–$3\frac{1}{4}$ inches (6·5–8 cm)
Wing-length: $7\frac{1}{2}$–$8\frac{1}{4}$ inches (18·4–20·8 cm)
Weight: $10\frac{1}{2}$–$12\frac{1}{2}$ ounces (300–350 g)
Diet: omnivorous
Number of eggs: 4
Incubation: 3 weeks

Adults
General colour brown, dark on the back with variegated markings. Belly lighter with narrow transverse bars. Large horizontal black bars across top of head and down neck. Black eyes situated high up and far back. Long, flexible bill; tip of upper mandible soft and mobile. Tail black, tip grey.

Young
Similar to adults but feet rather paler.

of the nesting cavity, supported on a callosity at the junction of the foot and lower leg. When their feathers have grown and are sufficiently firm, the young birds cling to the sides of the cavity and clamber towards the exit. Before they are three or four months old, shortly after they abandon the nest for good, their plumage changes completely but age can still be determined by the exceptional length of the first primary wing feather.

It is interesting to note that both the green woodpecker and the black woodpecker (*Dryocopus martius*) feed their young by regurgitating food whereas the representatives of the genus *Dendrocopos*, characterised by their black, white and red livery, offer the nestlings fresh prey from the bill.

Woodpeckers are very numerous throughout Europe and Asia. In addition to the species already cited mention should be made of the three-toed woodpecker (*Picoides tridactylus*) and the wryneck (*Jynx torquilla*). The latter is the most primitive member of the family and also the most specialised. Its bill is not all that strong for drilling bark nor are the tail feathers as stiff as those of other woodpeckers so that the bird normally perches across a branch instead of gripping the trunk. It is a skilful hunter of insects in mid-air but also feeds copiously on ants. It is also the only migratory representative of the woodpecker family in Eurasia.

The woodpeckers of America are even more abundant and varied, as well as being more brightly coloured than their Old World counterparts. They are divided into eight genera – *Dendrocopos* and *Picoides* (found too in Eurasia), *Colaptes, Hylatomus, Centurus, Melanerpes, Sphyrapicus* and *Campephilus*.

The mysterious woodcock

The woodcock (*Scolopax rusticola*) is a solitary bird, found in the most remote parts of the forest where the sun's rays hardly filter through the dense foliage and where the ground is heaped with a thick layer of wet, decomposing leaves. There is an aura of mystery about this bird, as there is with any species whose habits are insufficiently known and studied; and the wealth of traditional beliefs concerning its behaviour is from the scientific point of view of little or no value.

The contradictory information which has come so readily to hand probably arises from the fact that the woodcock is a highly individualistic bird and its behaviour so variable that observers have found it hard to come to any definite conclusions. There is, for example, much controversy as to whether it is a migratory species. Some birds certainly depart on long journeys, flying off into the valleys and towards warmer regions when winter arrives, while others are noted for being completely sedentary, continuing to rummage for worms even when the temperature falls below 0°C.

Except during the breeding season the woodcock spends its time deep in the forest undergrowth – sometimes on a bed of dead leaves or on a pile of logs, perhaps in a glade or at the side of a path, yet always merged with its surroundings by reason of its inconspicuous plumage and complete immobility. It becomes

Facing page: In March, as night falls, the male woodcock takes to the skies on his nuptial flight, slowly circling his territory and always following the same path. As he flies he emits raucous cries designed to attract the attention of the female who remains on the ground and responds to his summons with a gentle whistling sound.

The woodcock spends the day concealed in the woodland undergrowth. It is too timorous to run away from an enemy but often escapes notice by virtue of its cryptic coloration.

active at twilight, stretching, ruffling its feathers, and swinging its head from side to side. Its first concern is evidently to find some water for drinking and for cleaning its bill and feet. If this is near to hand the woodcock will proceed on foot, for although it is not as agile in this respect as the partridge it is able to walk and run without difficulty. Then, using its long, narrow, flexible bill, it rummages to a depth of two or three inches in humus and partially decayed vegetation or in the mud on the banks of a stream for its food. The woodcock's bill is indeed a marvellous tool, the upper mandible being especially mobile and sensitive, with delicate nerve endings. The bill is twisted this way and that until it comes up against a worm or insect larva. The bird then inhales, literally sucking up the prey which is already half swallowed as it is prised loose.

Some authors suggest that the jerky tread of the bird as it patrols its hunting grounds has a similar effect on the sensitive receptors of earthworms as does the pattering of raindrops and serves to draw them towards the surface. Apart from worms the woodcock also feeds on small molluscs, centipedes, spiders and, according to some authorities, seeds and other vegetation.

Because the bill forms such an efficient detecting device, the woodcock does not need to use its eyes–which are situated high up and towards the back of the head–in its search for food. The field of binocular vision is somewhat restricted in a forward direction but it can, without turning its head, see almost everything that is happening above or behind. In this way it can be continually on the alert, even when seeking food, and so avoid being surprised by a predator.

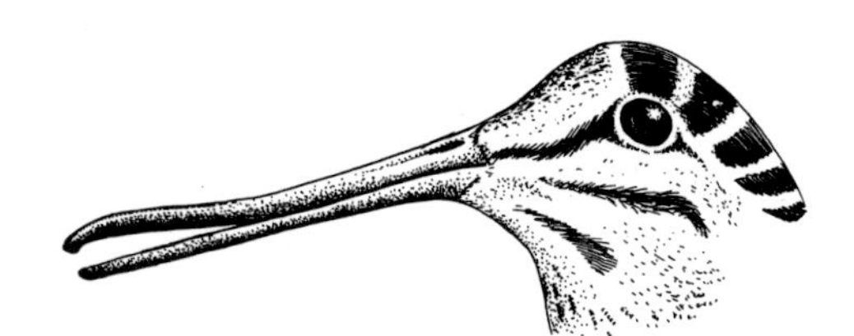

The woodcock uses its long bill to rummage in the soil and capture the worms which form the basis of its diet. The upper mandible is especially soft and sensitive and it is by means of constant probing that the bird manages to locate its prey which is then seized firmly, sucked up and swallowed.

March is the breeding season. At break of day, or later as the sun sets, when not even the hooting of the tawny owls breaks the silence of the forest, mysterious shapes detach themselves from the shadows and soar above the treetops. These are the male woodcocks, easily distinguished by the way their long bills point earthwards. Their flight paths at this season are regular and predictable (so much so that they are often easy targets for waiting hunters) and as they fly they emit mating calls in which harsh, muffled grunts alternate strangely with sharp, piercing cries. Down on the ground the females that are the cause of this furore respond to the males overhead with gentle whistling sounds. After a male alights near his intended partner there follows a passionate nuptial performance during which he droops and spreads his wings, opens his tail into a broad fan and slowly circles the female until she succumbs to his advances. Nevertheless, fidelity is not the rule. As soon as they have coupled, the female may accept the favours of another suitor while the male soars off on another nuptial flight.

The woodcock's nest is situated on the ground, in a hole clumsily covered with bits of grass and dead leaves, often at the foot of a tree. The female lays four eggs which are yellowish with brown and grey speckles. After three weeks' incubation the chicks hatch, their notable physical characteristic being a bill that is much shorter than that of the adults. Some reports, not scientifically confirmed, suggest that if the brood is endangered the mother will fly off with her youngsters, wedging them between her body and legs. In June the adults are ready to mate once more and the females are soon busy with a second brood.

The American subspecies is slightly smaller than the Eurasian race and has a more vivid plumage. Otherwise its behaviour is very similar to that of its Old World relative.

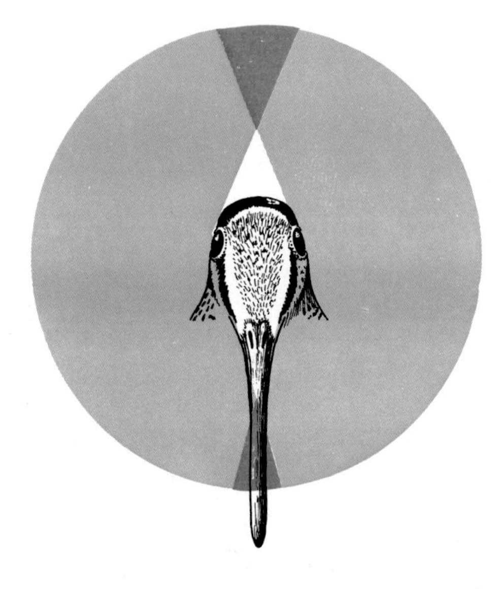

Because of the position of the eyes, high and far back in the head, the woodcock is able to rummage in the earth for food and at the same time keep a sharp look-out for predators approaching either from above or from the rear.

Uninvited guests—the cuckoos

On a spring day, especially around dawn or dusk, the clear, repetitive two-note call of the cuckoo (*Cuculus canorus*) can be heard from the depths of a wood—easily recognisable but extremely difficult to locate, for the bird does not often show itself. Although there are significant colour variations, notably in the female, the plumage of this medium-sized bird is a blend of grey, brown, black and white and in silhouette it looks much like a sparrowhawk, though its flight is considerably slower.

The migratory male cuckoos return to Europe in March and in early April, by which time they have settled in their territories and are prepared to defend them ferociously, they issue vocal invitations to the females who are just arriving from winter quarters. The solicited females signify consent by erecting their feathers and opening the tail in the form of a fan. Coupling usually takes place on the ground but sometimes high up on the main branch of a tree. As a general rule, according to surveys so far made, cuckoos do not form bond-pairs.

Having mated, the female cuckoo spends several days attentively watching the activities of a pair of small birds as they busily build their nest. When satisfied, she takes advantage of

Common cuckoo (*Cuculus canorus*)

Common cuckoo, rufous-brown form (female only)

Cuckoo Sedge warbler

Cuckoo Lesser whitethroat

The common cuckoo is a parasite, laying her eggs in the nests of other birds. The markings vary greatly, depending on the species selected as host and foster parent. Colour and pattern invariably match those of the host bird so closely that the suspicions of the latter are seldom aroused.

their temporary absence from the site to slip furtively into the nest, remove one of the eggs (later to be eaten) and replace it with one of her own. It is obvious that in order to dupe the rightful owners of the nest the cuckoo's egg should not clash in colour or pattern with those still there. It will in fact be astonishingly like the eggs already in the nest. There is no such thing as a typical cuckoo's egg – apart from standard structure and measurements. As far as colour and markings are concerned these blend very closely with those of the host species' own eggs.

No satisfactory explanation has been offered for this remarkable phenomenon although naturally there have been many theories on the subject. Edgar Chance believes that the female cuckoo normally parasitises the species that has reared her, laying eggs that are indistinguishable from those of this particular host. This would imply an acquired response originating from her earliest days in the foster parents' nest. Some statistics submitted by this ornithologist are of interest. Out of 61 eggs laid by the same cuckoo, 58 were deposited in the nests of pipits, matching them closely in colour and pattern.

It was the English zoologist Alfred Newton who first noted, in 1893, that cuckoos belonging to the same species (there are about forty in Eurasia) tend to separate into groups, each characterised by a certain type of egg. This raises the question of what kind of role the males play in reproductive behaviour. It would seem logical that a female 'specialising' in pipits should not mate with a male 'specialising' in sedge warblers, for the resultant eggs would be of an intermediate type and would immediately be detected and thrown out by the prospective

foster parents. Two answers to this fascinating question have been proposed. One suggests that the male cuckoo plays no part in determining the colour of the shell, that this is a sex-linked character. The other claims that the stimulus impelling the female to choose the nesting site, the song and the plumage of the bird to be parasitised is present in the male and that any individuals born to birds of the same host species would automatically tend to mate with one another. The former explanation seems the more plausible but there will be no positive conclusion until scientists manage to raise successive generations of cuckoos in captivity.

The initial task of getting rid of a host bird's egg and substituting her own sometimes presents problems. How, for example, does the cuckoo go about forcing her way into the little ball – a nest that has a small opening? It used to be suggested that the female cuckoo first laid her egg on the ground, then took it up in her bill and pushed it in from outside. The work of Edgar Chance, since confirmed by other observers, shows that she brings her cloaca up to the entrance hole, raises herself with outspread wings and tail against the opening and ejects the egg directly into the nest. It now seems probable that all cuckoos behave in approximately the same way when laying.

E. C. S. Baker has proved that out of 1,642 cuckoo's eggs laid in the nests of passerines that are traditionally selected as foster parents, only about 8 per cent were identified and thrown out by the host birds. What happens most frequently is that the pipit, sedge warbler or robin concerned returns to her nest, finds nothing to arouse her suspicions and settles down to incubate all

Some hours after hatching, the baby cuckoo instinctively begins to remove the eggs of its host, heaving them out of the nest one by one (*facing page*). Having asserted its right to be the sole occupant, the intruder is regularly fed by its foster parents, even when it has grown so large that there is no room in the nest. In autumn the fledgling departs for warmer climes.

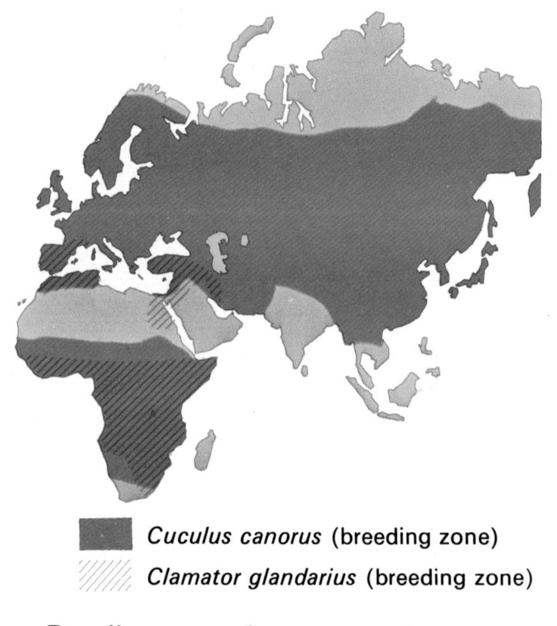

Breeding zones of common cuckoo (*Cuculus canorus*) and great spotted cuckoo (*Clamator glandarius*).

COMMON CUCKOO
(*Cuculus canorus*)

Class: Aves
Order: Cuculiformes
Family: Cuculidae
Total length: 13¼ inches (33 cm)
Wing-length: 8–9¼ inches (20–23 cm)
Wingspan: 23½–24¼ inches (59–61 cm)
Weight: 3–5 ounces (90–135 g)
Diet: principally insects
Number of eggs: 8–12 (one per nest)
Incubation: 12–13 days

Has similar appearance to sparrowhawk in flight. Long tail and pointed wings. Upper parts, head and nape greyish. Underparts striped brownish-black and white. Bill slightly curved. Feet yellow. Some females are rufous, with brown stripes on head and belly.

GREAT SPOTTED CUCKOO
(*Clamator glandarius*)

Class: Aves
Order: Cuculiformes
Family: Cuculidae
Total length: 15½–16¾ inches (38–42 cm)
Wing-length: 7¼–8¾ inches (18–22·2 cm)
Wingspan: 24 inches (60 cm)
Diet: insects
Number of eggs: 1–2
Incubation: 14 days

Grey crest; upper parts brown with white streaks. Tail grey with white borders. Underparts whitish. Young have no crest and back is darker than that of adults, belly yellowish or light brown.

the eggs. After twelve or thirteen days, usually a short while before its companions are due to be hatched, the baby cuckoo is born. It is naked, pink and blind, just a little bundle of flesh which seems to be incapable of coordinating its movements. Nevertheless, before it is much more than ten hours old it is transformed into a tiny devil with an inborn tendency to push away anything that happens to be in the vicinity. Huddled on the floor of the nest, it passes one wing underneath the nearest egg or fledgling and lifts it onto its back; then supported by its feet and the front of the head it slowly totters to the edge of the nest and heaves its burden overboard. It will repeat the operation as many times as may be necessary until it remains the sole occupant of the nest. Meanwhile the foster parents simply watch helplessly as the drama unfolds, often for several hours, impervious to the desperate cheeps of their own starving youngsters on the ground. The sight of the demanding open bill and orange-red throat of their uninvited guest seems to obsess them and for the next two or three weeks they make hundreds of journeys to satisfy the baby cuckoo's voracious appetite. Even when the youngster leaves the nest, having already become larger than its foster parents, the latter go on tending to its needs for a further three weeks.

Each cuckoo lays, on average, from eight to twelve eggs in different nests each season. Edgar Chance managed to obtain 25 eggs from the same bird by showing her a succession of birds building their nests–apparently a spur to reproduction.

The young cuckoos abandon their foster parents prior to leaving on their winter travels to warmer climes. But one case has been reported of a cuckoo that was unable to get out of the small entrance to the nest of a pair of wagtails that had reared it and which continued to be cared for until the middle of autumn when all other cuckoos had departed on their migrations.

As soon as they are capable of looking after themselves, the smaller insect-eating birds of the forest, as if intuitively sensing the danger posed to their kind by their intrusion, behave aggressively towards cuckoos whenever they meet them on the territories in which they were reared.

The adult cuckoo feeds on many insects, especially bristly caterpillars which other birds find impossible to digest and thus reject. Since many of these insects are harmful to crops, the cuckoo performs a useful service to man and deserves protection even though by human standards (never a good way to judge an animal) its behaviour may seem outrageously cruel.

The great spotted cuckoo (*Clamator glandarius*) appears early in March in the pine forests of southern Europe for it is not a bird of the deciduous forest but of Mediterranean woodlands. A close relative of the common cuckoo, its plumage is more like that of a magpie and its call particularly strident and varied. Nevertheless it too has parasitic breeding habits, with a partiality for nests of the crow family, other species seldom being victimised. Reliable information about its behaviour is hard to come by because of its furtive nature and unpredicatable movements. It seems, however, that the female lays her egg (sometimes two) when the owners of the nest are away and then flies off. The young great spotted cuckoo is not, however, as vicious as its cousin for it shows no desire to get rid of its foster brothers and sisters, settling down peacefully alongside them. The reason that the young cuckoo does not take the trouble to send its rivals packing is that it can use a far more subtle method of eventually getting things all its own way. Studies on the behaviour of this species have shown that the foster mothers feeding their broods are particularly stimulated by the bright pink gullets of their young as they open their beaks wide. But the mucous lining of the mouth of the baby cuckoo is even brighter and more conspicuous–so much so that the mother appears compelled to feed the intruder almost to the exclusion of her own brood. Within three weeks or so the well fed cuckoo is ready to leave the nest but still shrilly demands food from the adults. They are so busy fetching and carrying outside the nest that they completely neglect their own half-starved progeny, dooming them to certain death.

In areas where both cuckoos and magpies live together, the latter are often too numerous, destroying the eggs and nestlings of other species, so that this apparently callous behaviour helps to control the magpie population. The cuckoo also feeds on caterpillars and is thus beneficial to forest growth.

The great spotted cuckoo is an inhabitant of the Mediterranean woodlands and with its black and white plumage and long tail is often taken for a magpie. Like the common cuckoo it lays its eggs in the nests of other birds, particularly Corvidae.

The Coraciiformes are divided into four suborders. The Coracii are the rollers, here represented (*left*) by the Abyssinian roller (*Coracias abyssinica*). The Meropes are the bee-eaters, belonging to the one family Meropidae. The birds in the photograph (*right*) are European bee-eaters (*Merops apiaster*). (see also p. 275 bottom right).

ORDER: Coraciiformes

The birds of the order Coraciiformes have a widespread distribution in many parts of the world, inhabiting most tropical and warm regions.

Among a number of characteristic anatomical details which these birds have in common is the desmognathous structure of the palate (the palatine apophyses of the jaw being joined along a central line).

The foot is anisodactylous, with three toes facing the front and one towards the rear (the exception being the cuckoo roller). Two or three of the toes are joined for part of their length, but in the case of the rollers the outer toe moves freely. The claws thus function rather like pincers. The sole of the foot is well developed and the claw of the middle toe is generally larger than that of the big toe, except with the hoopoes where the claw of the rear toe is the best developed.

These birds are often gregarious and sometimes migratory. Although some (notably certain hornbills and the rollers) are ground birds, most are essentially tree-dwellers. They vary considerably in size. The smallest representative of the order is the red-throated bee-eater of Africa, which is about 8 inches long. The largest birds are the hornbills, some of which measure almost 5 feet long.

The wings may either be short and rounded or long and pointed. Flight patterns too are variable, that of the bee-eaters being graceful but erratic, that of the kingfishers rapid and sustained, that of the hoopoes undulating, with butterfly-like wing movements. Hornbills, in the course of their lumbering flight, make a noise rather like steam escaping from a valve (similar to the sound emitted by toucans). But from the point of view of acrobatics the most spectacular displays are put on by the various rollers.

The tail, comprising 10–12 rectrices, is variable in length. In the case of the motmots of Central and South America it is very large, some inches longer than the body. In certain species of Coraciiformes the tail feathers are trimmed so that the tip is enlarged like a racquet.

Plumage is usually sparse but with brilliant, often contrasting colours. An oily substance prevents the feathers of kingfishers staying wet after they have dived for fish.

The stout, strong bill varies in size but that of the hornbills is disproportionately large, sometimes with a horny growth or casque. Motmots have saw-toothed edges to the bill which help to dismember insects.

Except in rare circumstances birds of this order nest in natural cavities or in dead trees. Some dig underground galleries running in a horizontal direction and terminating in a small chamber where the eggs are laid. The tunnels of certain bee-eaters may be up to 10 feet long, the digging of which occupies both male and female for many hours. The circular entrance hole to the nest is just large enough to accommodate the bird's body so that each adult has to take turns in going in and out. The main purpose is of course to guard against raptors and other enemies but it cannot always prevent a snake from gliding in to steal eggs or attack chicks.

The number of eggs, depending on species, is 2–8 and these are shining white, exceptionally transparent or pink. Both parents participate in the four-weeks' incubation. In the case of the hornbill, however, only the female incubates the eggs, sealed in her nest and fed by the male.

The nestlings are nidicolous. Although born naked they soon show signs of little sheaths out of which the feathers grow and which disappear when the fledglings are ready to leave the nest. The plumage is normally the same colour as that of the adults.

Food consists principally of insects, small mammals, amphibians and fishes. The hornbills are omnivores.

The hunting technique usually follows a set pattern. The birds normally perch on a tree or hover around for prey. After swooping on the victim, which they either stun or kill outright, they carry it back to the roost. When kingfishers catch a fish they fly back to their rock or branch, kill the prey, toss it in the air and swallow it head-first. Some rollers join other birds in hunting forays.

Occupation and defence of territory is an important feature of behaviour. Pairs of kingfishers, for example, will take up residence on a stretch of ground along a river bank (the size will vary in accordance with the amount of prey available) and exclude all rivals.

The birds also defend themselves boldly against enemies. Some hornbills and rollers close ranks to confront an aggressive intruder or emit such raucous cries that the raptor eventually flies off. Young and female hoopoes have an additional weapon at their disposal—a gland at the base of the tail which exudes a musky secretion.

There are four suborders of Coraciiformes. The Halcyones include the Alcedinidae (kingfishers, with a worldwide distribution), the Todidae (todies, from the Caribbean islands), and the Momotidae (motmots, inhabitants of tropical America). The Meropes comprise the one family Meropidae (bee-eaters of the Old World). The five families belonging to the suborder Coracii are the Brachypteraciidae (ground-rollers of Madagascar), the Coraciidae (rollers, from the tropical regions of the Old World), the Leptosomatidae (one species only, the cuckoo roller from Madagascar), the Upupidae (hoopoes, found in Europe, Asia and Africa), and the Phoeniculidae (wood hoopoes, from Africa). The suborder Bucerotes comprises the single family Bucerotidae (hornbills, from south-east Asia, the East Indies and Africa).

Above : The grey hornbill (*Tockus nasutus*), a representative of the suborder Bucerotes. *Below :* The lesser pied kingfisher (*Ceryle rudis*), one of the Halcyones.

ORDER: Piciformes

In many woodland and forest regions, particularly at the hottest time of the day, a steady, hollow tapping noise, sometimes as loud as the sound of a hammer striking an anvil, reveals the presence of a woodpecker, probably the best known bird belonging to the order Piciformes.

One of the most important physical characteristics of the numerous Piciformes is the foot structure, perfectly adapted as it is for climbing. The foot is zygodactylous (yoke-toed), with two toes facing to the front and two towards the rear. The first, second and fourth toes are all activated by the flexor muscle of the first toe but the third toe is independently controlled by the usual flexor muscle of the toe.

This arrangement of toes (which is shared by parrots and cuckoos) is indispensable for taking a firm grip on the tree surface, even when the trunk is vertical. In some woodpeckers the first toe is very small or not present at all (so that there are two toes in front and only one at the back). In all cases, however, the sharp curved claws sink naturally into the rough or semi-smooth bark.

The Piciformes include many climbing birds which resemble some members of the Coraciiformes both in behaviour and in their wide global distribution. They are in fact found in every part of the world with the exception of Australia, New Zealand, the Oceanic islands and Madagascar.

Another interesting feature of many of these birds is the unusual structure of the hyoid bone at the root of the tongue. This is so large that it encircles the skull. Attached to it are the powerful muscles controlling the protractile tongue.

This is a characteristic of the woodpeckers which have in addition a skull so structured that it can sustain the continuous drumming movements of the head and remain unaffected by the vibrations caused by the incessant blows of the bill against the tree. The tongue is long, narrow and extremely mobile, with tiny backward-pointing horny barbs at the tip which not only help to perforate the bark but also function as miniature harpoons for capturing insects. Woodpeckers also possess special salivary glands, parts of which secrete a watery substance and others a sticky, glue-like fluid which covers the surface of the tongue and serves to trap insects and larvae. The tongue of the toucan, however, is long and flattened, the edges being fringed with bristles and the tip tapered. This is in conformity with the fruit diet of the bird.

Many toucans have enormous bills, outmeasuring those of all other birds with the exception of certain hornbills. A toucan whose body is some 20 inches long may have a bill that measures about 24 inches. Although it looks completely out of proportion this bill is not only strong but also fairly light-weight, comprised as it is of a network of spongy cellular fibres.

The bills of other Piciformes are generally quite large with a blade- or chisel-like tip. A few are slender and curved, even blunted. But no matter what the shape they are powerful enough to pierce wood with comparative ease, not only with a view to dislodging insects but also to building a nest.

Although these birds will sometimes nest in a hole or even in the steep banks of a stream, they frequently use a tree trunk as a nesting site, drilling a rounded passage which is expanded at the base to hold the eggs. Like those of the Coraciiformes, the shells are pure white since there is no risk of their being seen from the outside. The number is 2–8, according to the species, and incubation lasts 40 days at most, duties being shared by male and female. The young, which are nidicolous, have a curious habit, similar to that of young kingfishers and hornbills, of folding their tail over the back.

Many Piciformes lay their eggs in the nests of other birds belonging to the same order. Among these are the honeyguides. The females of these species deposit their eggs in the nests of woodpeckers or jacamars. As soon as the baby is hatched it destroys the nestlings of the host bird by pecking them with the sharply hooked tip of the bill. When the nest becomes too small for comfort the young honeyguide proceeds to enlarge it.

The beak of the nestling is unique. The tip of both upper and lower mandible ends in a sharp hook. These hooks curve towards each other, turning slightly inwards towards the cavity of the mouth. With them the nestling honeyguide wounds the legitimate nestlings, sometimes throwing them out of the nest.

Although they form a heterogeneous group with certain important resemblances which give rise to their being classified in the same order, the Piciformes have other common features. The plumage, for example, is stiff, fairly sparse and often brightly coloured. The jacamars and some of the toucans are among the most spectacular of all birds. Others are drabber, without strongly contrasting colours, and individuals of both sexes are often alike in appearance. Coloration is in many cases related to camouflage, blending with plants, so deceiving both potential prey and predators.

Insects form the basis of the diet of most of these birds and hunting methods are similar to those employed by the Coraciiformes. They settle on a favourite branch, swoop suddenly on their insect prey, carry it back to the roost, then kill and devour it. The birds often show a particular attachment for the same branch and the various puffbirds, for example, may remain in one spot for years on end. Some jacamars use an ambushing technique to pursue their prey or glide above their victims in the manner of certain raptors.

The toucans are exceptional in being frugivores, feeding on a wide range of berries and fruits.

The typical feeding methods of the insect-eating species, including the woodpeckers, are most ingenious. The birds hop up and down bare tree trunks, pecking tiny holes in the bark and using their barbed tongue to extract xylophagous insects and larvae from the cavities.

The wings of Piciformes are short and rounded – a structure characteristic of all birds that are poor fliers and also sedentary.

The rectrices or tail feathers of woodpeckers are not only extremely strong but also very flexible, enabling the birds to use the tail as a support when climbing.

Modern classification, based on Peters, divides the Piciformes into two suborders, Galbulae and Pici.

Five of the six families comprising the order belong to the Galbulae. The Galbulidae are the jacamars – slender, colourful, insect-eating birds from Central and South America which usually build their nest in a bank or hillside.

The Bucconidae are the puffbirds, closely related to the jacamars and also exclusive inhabitants of the New World. They too feed on insects and nest in burrows at ground level. Inconspicuous both in behaviour and appearance, they are sometimes nicknamed 'nunbirds'.

The barbets belonging to the family Capitonidae, closely related to toucans and honeyguides, are found both in the Old and the New World. Their name is derived from the long bristles on the bill which sometimes resemble a small beard. Most live in Africa but there is one Asiatic genus (*Megalaima*).

The honeyguides of the family Indicatoridae have an extremely wide distribution range throughout Africa and southern Asia as far east as Sumatra and Borneo.

The Ramphastidae are the toucans – inhabitants of Central and South America, one genus (*Ramphastos*) having given its name to the family.

The suborder Pici is comprised of the one family Picidae – the woodpeckers. There are 210 species, most of them in the Old World. The two principal genera are *Dendrocopos* (the pied woodpeckers) and *Picus* (the green and grey-headed woodpeckers).

The remarkably long hyoid bone of Piciformes encircles the skull. By contracting the powerful muscles attached to the bone the birds are able to 'harpoon' insects and larvae in the bark of trees with their protractile, barbed tongue.

Facing page : The toucan (*above*) belongs to the Galbulae, one of the two suborders of the Piciformes. The woodpeckers belong to the Pici, the one pictured here (*below*) being the grey-headed woodpecker (*Picus canus*).

ORDER: Cuculiformes

For some time the Cuculiformes were classified together with the Piciformes, principally because both cuckoos and woodpeckers have zygodactylous feet (two toes to the front and two to the rear). More detailed study, especially concerning the behaviour of cuckoos, convinced ornithologists that a separate order should be established for the latter species. Some authors place the Cuculiformes midway between the Psittaciformes (parrots) and Piciformes (woodpeckers); Alexander Wetmore sees them as occupying an intermediate position between the Columbiformes (pigeons) and Strigiformes (owls).

The Cuculiformes–birds whose bill frequently lacks a cere and whose tongue is small and flat–are divided into two dissimilar families.

The Cuculidae comprise the cuckoos proper, the coucals and the roadrunners. They are inhabitants of both the Old and New World. The Musophagidae include the touracos from Africa.

The relationship between these two families is so tenuous that many authorities prefer to dissociate them completely. But experiments related to the protein content of egg albumen and certain anatomical details, such as the structure of the feet, justify the two families being grouped together.

The Cuculidae feed chiefly on insects and their sizes range between that of a sparrow and a crow. Most of them are tree-dwellers but a few have terrestrial habits.

The females belonging to the numerous genera of this family habitually lay their eggs in the nests of other species, ensuring that their young are reared by foster parents. The majority of these parasitic cuckoos belong to the subfamily Cuculinae, comprising 16 genera and 47 Old World species. Best known is the common cuckoo (*Cuculus canorus*) but among other notable species are the great spotted cuckoo, the coucal of tropical Africa, the emerald cuckoo of South Africa, the Indian koel and the golden-bronze cuckoo which breeds in New Zealand and migrates from April to September to the Solomon Islands.

The cuckoos which do not have parasitic habits make up the majority of the other five subfamilies of Cuculidae. They include the yellow-billed and black-billed cuckoos which build flimsy nests of grass and twigs in trees and lay 2–6 eggs which are usually bluish but may be variously coloured and marked, either uniform or speckled. Both parents collaborate in incubating the eggs and feeding the young.

The anis, black cuckoos from tropical America, are very gregarious and construct large nests in which several females lay their eggs.

The notable feature of the subfamily Neomorphinae is that five of the six constituent genera are found in the New World and only one in tropical Asia. Two genera (three species) are parasitic, the rest are not. The best known member of this subfamily is the roadrunner (*Geococcyx californianus*) which feeds on small reptiles.

The touracos belonging to the family Musophagidae, as already mentioned, differ from the Cuculidae and in the view of some authors more closely resemble the Galliformes. There are five genera and eighteen species, all of them residents of Africa. Unlike the Cuculidae they feed chiefly on fruit and particularly plantains (an alternative name is 'plantain-eater'). The bill is usually strong and curved but shape and measurements are variable. Most of them have a prominent crest and are vividly coloured–mainly green, red and blue. In some species the feathers of head and neck lack barbules. The tail is long and the birds climb and run better than they fly.

Touracos live deep in the forest and build a flimsy platform-type nest of twigs. The two eggs are usually white and are incubated by both parents. The adults regurgitate food for their young.

Facing page: The two families making up the Cuculiformes–Cuculidae (cuckoos) and Musophagidae (touracos)–have little superficial resemblance to each other although they do possess certain common anatomical features. Livingstone's touraco (*Tauraco livingstonii*) is one of the several spectacularly plumaged species, all of which are found in Africa.

CHAPTER 14

The bears: giants of wood and mountain

At one time bears were familiar inhabitants of the mountains and forests of Europe, Asia and North America, roaming regions where there was a reasonably good supply of water and where summers were not too impossibly hot. Nowadays the last representatives of this wide-flung family (Ursidae)–not of course taking into account the polar bears–are restricted to a few remote hill and mountain districts and unfrequented forests.

The bear's relationship with man goes back to prehistoric times. Images of the animal have been found engraved on cave walls dating back to the Quaternary period. The hunting scenes depicted so vividly by these earliest artists, as well as the discovery of bears' bones and neatly stacked skulls in sites close to those containing human remains, suggest that the giant cave bear (*Ursus spelaeus*) was not only hunted by Palaeolithic man but was also a cult figure in magical rites.

Even today the primitive Ainu people of Japan, living in the Sakhalin Islands and Hokkaido, worship the bear as the incarnation of one of their gods, trapping and rearing the animal and finally sacrificing it as the climax of a solemn religious ritual.

The fact that the bear is capable of standing upright on its hind feet and of shambling forwards in a clumsy approximation of human gait may well account in part for the fascination it has traditionally exercised on the primitive mind but, superstition apart, the bear is acknowledged to be an unusually intelligent and adaptable animal. In a world where monkeys did not exist one can understand how it could arouse contradictory feelings of fear and veneration. Nor is it surprising that it should have featured prominently in folklore, myth and heraldic art.

Man's predatory instincts, however, gradually gained the upper hand and the attitude of later hunters, whether armed with

Facing page: Brown bears are not nearly so aggressive and harmful as is commonly believed, but this has not prevented these peacefully inclined animals from being hunted relentlessly. Once widely distributed throughout the Holarctic region the remaining individuals in the wild have taken refuge in remote mountain forests. Most brown bears, however, are to be found today in reserves, especially in Europe.

The brown bear is often seen in a seated position, which it adopts for a variety of activities, especially for picking and eating berries in the undergrowth or for feeding on other ground vegetation.

Facing page: Persecution of the brown bear reached its peak in Europe, resulting not only in a drastic reduction of numbers but also in a complete reversal of its behaviour pattern. Whereas the bear had originally been freely active by day it gradually adopted nocturnal habits, avoiding human presence whenever possible. In the wild daytime activities are kept to a minimum.

bows and arrows or telescopically sighted rifles, was far less equivocal. Killing a bear for its flesh can be justified but slaughter for its own sake is cowardly and despicable. The later chapters in the history of man versus bear make sorry reading.

No two the same

Although zoologists talk glibly of the 'brown bear' this description tends to be misleading for the colour of the animal is extremely variable. Within the same fairly restricted locality no two bears are precisely alike even if they nominally belong to the same species. The colour of the fur may range from near-yellow or cream through every nuance of brown to jet black; and those hues may be subtly modified in countless ways with blended shades of russet, cinnamon and silver-grey. To complicate matters still further some bears have light brown fur in spring which turns appreciably darker in autumn after the moult.

There are enormous discrepancies in weight as well, and since these tend to be more clearly marked in regions fairly widely separated from one another it was principally on the basis of such differences that zoologists were tempted to identify a number of separate species of brown bear. Thus only a few years ago it was still accepted that the brown bear of the Pyrenees–the males of which may weigh about 650 lb–did not belong to the same species as the Manchurian brown bear which may be up to 250 lb heavier; and since both of these were greatly outweighed by the male grizzly bears of North America (the largest of which are over 1,400 lb) and by the gigantic Kodiak bears of Alaska which tip the scales at about 300 lb more, it seemed reasonable to suppose that the latter were likewise distinct species.

Today Marcel Couturier and others have shown that there is only one species of brown bear, *Ursus arctos*, with a number of racial variants which have come about as a result of geographical isolation and in some cases for diverse ecological reasons. Other zoologists prefer to distinguish several species and subspecies. The American black bear is now classified in a separate genus as *Euarctos americanus*, though even this is not universally accepted. In any event brown bears once formed a flourishing community inhabiting the entire Holarctic region.

The physical differences between these races, subspecies or species stem from a wide distribution as well as from their ability to adapt to a wide range of surroundings, from sea level to altitudes of several thousand feet; and the principal reason for this distribution and adaptability is that the animal is an omnivore which can take the fullest advantage of all the food resources of a particular environment. It is worth examining this eclectic food range in some detail since this is the brown bear's most strikingly original feature.

Habitat and diet

Not so long ago the brown bear was found in the Vosges, the Juras, the Massif Central and the Alps. Nowadays the last survivors are confined to the Pyrenees and the Cantabrian Mountains.

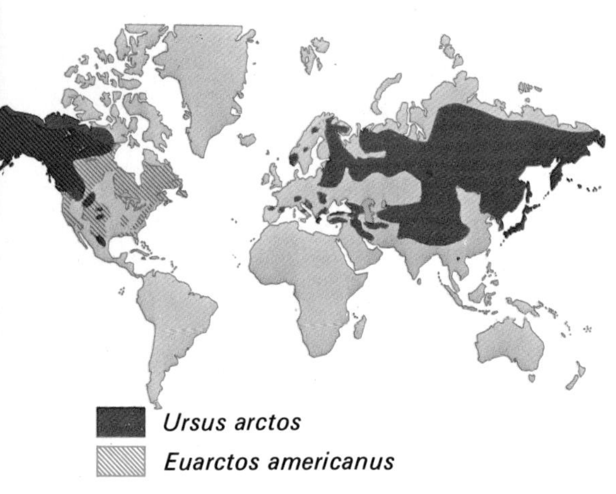

Geographical distribution of the brown bear (*Ursus arctos*) and the American black bear (*Euarctos americanus*).

BROWN BEAR
(*Ursus arctos*)

Class: Mammalia
Order: Carnivora
Family: Ursidae
Total length: 68–80 inches (170–200 cm)–Pyrenees; 100 inches (250 cm)–Carpathians; over 120 inches (300 cm)–Alaska; 120 inches (300 cm)–Kodiak I.
Length of tail: 4½–8 inches (11–20 cm)
Height to shoulder: 36–40 inches (90–100 cm) Pyrenees; up to 48 inches (120 cm)–Carpathians
Weight: 175–660 lb (80–300 kg)–Pyrenees; up to 1,440 lb (655 kg)–Kamchatka; 1,700 lb (780 kg)–Kodiak I.
Diet: omnivorous
Gestation: 180–250 days
Number of young: 1–3

Adults
Heavy body, short legs, large head, small eyes, short and rounded ears. Colour very variable, from pale yellow and silver-grey through brown to deep black–often blended with other shades (red, cinnamon, fawn, chestnut, black and grey). Females are less robust.

Young
Newborn cubs are tiny, weighing less than 1 lb, blind and toothless. Only the claws of the forefeet are well developed with sparse hairs, pale at first but soon becoming darker. When they leave the den their thick cover of fur is chestnut or chestnut-grey, sometimes black, with white or whitish throat. At six months the fur is paler and at sixteen months they resemble adults.

The brown bear cannot tolerate excessive heat or drought and must have access to water for drinking, with cool places for resting. Glaring light is hard on the animal's eyes (its vision is not good at the best of times) which accounts for its habit of loping off into the woods and undergrowth in the middle of the day. It rummages in moss for larvae and worms and has a partiality for soft, wettish soil thickly covered with humus where it can easily scoop out bulbs, tubers and roots. It often seeks refuge on rough hilly ground where there is little risk of human interference. Since it has no natural enemies interaction with other zoological groups plays little part in choice of territory, although the presence of ungulates and the proximity of beehives may sometimes prove decisive. But the principal determining factor is the composition of the plant cover–the most favoured habitat being virgin forest.

Couturier has provided detailed information about the behaviour of the brown bear in the Pyrenees, normally to be found at an altitude of around 5,000 feet but sometimes at heights of up to 9,000 feet. The vegetation here is characteristically that of the deciduous or mixed forest, including pine, beech, oak, Austrian oak, chestnut and birch. The undergrowth is made up of elder, holly, yew, hazel, honeysuckle, willow and ivy. Sheltered by this canopy of leaves are plants which provide the bear with a rich variety of food–currants, raspberries, pears, bilberries, bearberries and strawberries.

Within a region which on the surface appears to offer a standard range of food possibilities, an individual bear obviously takes account of minor variations to settle in one area rather than another. Even an ecologist may be hard put to it to detect any significant reason for such a choice.

The home range of a brown bear may extend for 8–10 miles but in regions unexplored by man it can be much more extensive. Yet the animal does not defend its domain with the same jealous ferocity and determination as other territorial species. Two bears may frequently share the same hunting grounds. The scratches which are often found on tree trunks cannot therefore be regarded as frontier signs nor even as landmarks, for the bear has a wonderful memory for topographical detail.

Most animals of savannah and steppe concentrate on a somewhat uniform diet but the forest-dwelling bear has such an enormous variety of food at its disposal–fruit, bulbs, mushrooms, insects, small mammals and birds–that it can afford to be selective. So great is the choice that despite its proverbial wisdom the animal may be tempted to eat more than it can manage. This often happens when it finds a supply of small, succulent berries which are rather difficult to pick. To satisfy its hunger quickly the bear will grab hold of enormous handfuls. Fruits and other vegetable substances are special favourites and this is linked with the animal's dentition. The teeth structure was originally that of a typical carnivore but has gradually been modified in the course of evolution to that of an omnivore.

The actual composition of the brown bear's daily meal varies from season to season. When green shoots are scarce the animal concentrates on fresh meat. Although its hunting methods are

typically opportunistic, it is certain that the parents teach their young how to go about finding food, if only by example, and that early instruction and imitation determine their later partiality for one kind of food in preference to another. Acquiring the complicated techniques of capturing live prey is important for bears that specialise in hunting birds and mammals as well as for those more skilled in catching fish.

The bear makes use of its claws, rather as we do fingers, for grasping food and carrying it to the mouth. It is interesting to note that some individuals are right-handed, others left-handed. The curving claws of the forefeet are obviously extremely powerful but what is astonishing is their flexibility. The dexterity with which a bear uses its paws to dislodge stones concealing worms and insects, to rifle honey from the nests of wild bees, to trap a trout or a salmon and if need be to fell a sheep or a calf, is quite remarkable.

In spring the brown bears of the forest may include in their diet earthnuts (*Conopodium majus*) and tubers, roots and rhizomes of such plants as the ramson or broad-leaved garlic (*Allium ursinum*), saffron (*Crocus sativus*), broomrape (*Orobanche*), trefoil (*Trifolium*) and cuckoo-pint (*Arum maculatum*). Fungi are greedily devoured–particularly morels (*Morchella*), milkcaps (*Lactarius*), boletus (*Boletus*) and puffballs (*Lycoperdon*)–as well as truffles. Among many types of fruit consumed with special

Born in the winter, the cubs are affectionately tended by the mother for a year and may possibly occasionally spend the second winter with her. By the following year, when she is carrying another litter, they have asserted their independence.

AMERICAN BLACK BEAR
(*Euarctos americanus*)

Class: Mammalia
Order: Carnivora
Family: Ursidae
Length of head and body: 60½–72½ inches (152–182 cm)
Height to shoulder: 24–36½ inches (61–92 cm)
Weight: 265–330 lb (120–150 kg)
Diet: omnivorous
Gestation: 100–210 days
Number of young: 1–4

Similar in appearance to the brown bear but with longer muzzle and shorter claws. Colour is variable–black, bluish, cinnamon or almost white, but muzzle always chestnut.

Facing page: Few animals enjoy such a broadly-based diet as the omnivorous brown bears. Though fruits, berries and other forms of vegetation make up a large proportion of their food there are times when carnivorous instincts prevail. In some regions they may specialise in hunting birds; elsewhere they will prey on domestic flocks and herds grazing in mountain pastures.

relish – and the bears may even climb the branches and risk losing their balance to get at them – are the crab apple (*Malus pumilus*), wild pear (*Pyrus communis*), wild cherry (*Prunus avium*), sour cherry (*Prunus cerasus*), medlar (*Mespilus germanica*), grape (*Vitis vinifera*), sloe (*Prunus spinosa*) and berries of the holly (*Ilex aquifolium*) and juniper (*Juniperus communis*).

In August and September the bears gather strawberries, raspberries and, to a lesser degree, pears. Later in the year they eat currants, barberries (*Berberis vulgaris*), mullein (*Verbascum*) and bearberries (*Arctostaphylos uva-ursi*). Other delicacies include the hips of dog roses, in spite of the hairy carpels which may bring on fits of coughing. From October onwards the animals come down from the mountains to look for areas where there are plenty of acorns, chestnuts and the like. To a lesser extent they also feed on walnuts and hazelnuts, while a particularly good crop of beechmast may induce them to delay hibernation. This long list of edible substances is rounded off with the acid fruits of the rowan (*Sorbus aucuparia*), the stems of umbelliferous plants such as the wild carrot (*Daucus carota*) and wild angelica (*Angelica sylvestris*).

It is obvious from this survey of food items that carbohydrates play an important role in these animals' nutrition. They are assimilated and stored in the body as fats, enabling the bears to live on these reserves during the winter.

As far as meat is concerned this consists principally of animal protein. Lambs, for example, are cleanly skinned before being eaten and are more often killed than goats which are rather more difficult to capture. Cattle, horses and asses are occasionally attacked. Pigs are special favourites, including wild boars (both the young and the odd adult which does not weigh more than 60 lb or thereabouts); and other victims may include a chamois or a sick or injured roe deer. Among smaller mammals hunted are squirrels, insectivores such as moles and shrews, and rodents (notably garden and fat dormice). Birds such as wood pigeons and grouse are first plucked of their feathers. Fresh meat is normally preferred but, if starving, bears will not shun carrion.

The diet is sometimes completed with fishes (trout and, in North America, salmon), lizards, frogs, snails and ants. Honey and eggs are choice delicacies, if available. Water is lapped up with the tongue, quantities being moderate, except in summer.

Brown bears are not afraid to enter water and are skilled at catching fish. In North America they patrol river banks as salmon head upstream during the spring to their spawning grounds, catching them with the mouth.

Daily life of the bear

It has been scientifically established that although the bear will prowl by night, chiefly with a view to avoiding contact with humans, in many sparsely inhabited areas such as Siberia, Kamchatka, Manchuria and Alaska, it retains diurnal habits. But in the mountainous regions of western Europe it hardly ever moves around by day and only an occasional glimpse of it will be obtained. Daily activity normally begins at dusk and in spring the bear will seek food continuously until dawn breaks. Only a very dark night will be chosen for an attack on livestock, and in order to retrace its steps to the scene of a crime the animal will often use the tracks of the shepherds themselves and their flocks,

Although they look rough and clumsy the forepaws of the brown bear, furnished with sharp claws, are not only extremely powerful but are also capable of performing a variety of delicate actions, some of which are shown in these drawings.

together with paths that often lead through dense undergrowth.

Weather, particularly in extreme forms, has a considerable influence on behaviour. Heat slows down body activity whereas cold (at times when the animal is in winter sleep) stimulates it. Wind, rain and snow seem to be treated with equal indifference.

As might be expected, daily behaviour varies with the season. At the end of the long winter sleep the animal is emaciated as a result of its fast and spends most of the time feeding. At this period the diet is chiefly vegetarian and in the Pyrenees consists in the main of earthnuts. In its weakened condition the bear is in no fit state to hunt. In late spring the natural products of the forest provide plenty of food and it is only in early summer when flocks and herds are led up to the high pastures to graze that the bear turns in earnest to hunting. Animal victims are not necessarily consumed at a single sitting and uneaten parts are often hidden under a heap of branches to await a return visit some forty-eight hours later. Summer is also the mating season. By this time the bear has put on weight and is less preoccupied with food. It is especially timid and apprehensive and although one might expect the contrary it is less often seen now than at any other time of year.

Moulting is completed by August when the woodland fruits are beginning to ripen. Towards the end of September food includes enormous quantities of acorns and beechmast and in October the bear roams the valleys in search of chestnuts. This continuous quest for vegetable substances entails a good deal of moving around and sometimes compels the bear to take unusual risks. Thus from November onwards it may prowl by day. Towards mid-December almost all the bears go into winter dormancy, which is not true hibernation.

Winter sleep

For the greater part of the year the brown bear makes use of a number of lairs, taking no great pains to make them comfortable and frequently moving home. But when winter comes there will be only one refuge and this will be carefully selected and meticulously prepared. In the Pyrenees it generally takes the form of a natural cave (at heights of 3,500–6,000 feet) facing the south, or it may be situated in an area of dense undergrowth deep in the forest. Should the entrance to the cave be too large the animal will build a kind of partition consisting of heaped branches of pine or beech, carefully sealing any gaps with moss and leaving only one small aperture. Sometimes it will block up the entrance completely, trampling the vegetation and replacing it every time it goes in or out.

The bear then scoops a narrow depression in the floor of the driest part of the cave and installs a kind of litter made of grass and dead leaves, so thick that when the animal is lying down its body is hard to detect.

If it is unable to find a suitable cave the bear will seek shelter in the hollow of a tree, covering the entrance with dry branches. After a heavy snowfall the animal is completely enclosed and protected inside its 'igloo'.

It is hard to say exactly when the bear retires for the winter for the date inevitably varies according to locality and climatic conditions in any particular year. In the Pyrenees winter dormancy usually lasts from mid-December until mid-March; but in lowland valleys it may not commence till the end of December. At the start of this long rest period the bear will continue to leave the lair from time to time to defecate and urinate, in conformity with its characteristically clean habits. Excretory activity diminishes as the animal gradually eats less and less but it is still compelled to make the occasional appearance out of doors to empty its bladder.

During the ensuing winter months all the body functions are in fact slowed down, especially the rate of heartbeat and breathing. The body temperature also falls although sensory perceptions are not affected. The most characteristic feature of this lethargic condition is the manner in which the body uses up its reserves of fats–vital for survival in view of the fact that the animal is now dispensing with food.

Since the bear's sleep is periodically interrupted by waking periods it is not strictly correct to refer to it as a state of catalepsy (characterised by suspension of all sensation, muscular rigidity and so forth), as happens in the case of the North American poorwill. A particularly fine, sunny day may even tempt the animal out of doors. Some authors have therefore come to the

Following pages: Although brown bears are normally solitary animals groups may sometimes be seen engaged in common activity. Such is the case in North America when a number of adult bears go fishing for salmon. There are no signs of mutual hostility; indeed each bear pretends to ignore its neighbour.

The brown bear must have regular access to fresh water. Although it only drinks in moderation it bathes frequently.

The brown bear of the Pyrenees is a shy animal and very wary of humans. Although eyesight is mediocre hearing and scenting powers are keen so that it can detect an intruder without difficulty.

conclusion that certain individuals do not become dormant at all, this belief being supported by observations of bears which have been disturbed during the winter by hunters and which have evidently abandoned their lair to seek shelter elsewhere.

Strength and mobility

The various races or subspecies of *Ursus arctos* all possess extraordinary physical aptitudes. It is obvious at a glance that they are uncommonly strong animals and the muscles of the neck, back and forefeet are especially powerful. But for such a heavy animal the brown bear is remarkably agile as well. It is a plantigrade, walking, like man, on the soles of the feet. Incidentally, the skin of the sole moults in March–the time when the bear is beginning to be active–and this is why it continuously licks its feet at this season.

Gait and speed are very variable, according to circumstances, with the back feet touching the ground ahead of the forefeet when the pace is fast. The bear does not trot but rather ambles, in the manner of a camel, this form of locomotion being just as speedy. It seldom resorts to a gallop and in fact only breaks into a run if it is attacking prey or fleeing from a man. But although its gait is somewhat broken and disorderly–not to be compared, for example, with the sustained pace of a horse–the bear can, if required, cover short distances at rates of up to 30 miles per hour–much faster than any human athlete.

The bear will often adopt the familiar squatting-on-haunches posture of the dog and will sometimes shuffle about on its rump to gather fruits and berries. Although it is capable of standing upright it seldom adopts this bipedal position in the wild; should it do so it is usually to get a better view of the surroundings, not to launch an attack. Nor does this exhaust its methods of getting about for it shows considerable skill in clambering over obstacles, climbing trees, swimming and diving.

Sense and intelligence

Zoologists seem to be agreed that the brown bear, no matter where it is found, has pretty poor eyesight. The eyes are in fact rather small and closely placed together, providing a wide field of binocular vision within a radius of approximately 80 yards but not equipping the animal too well for long-distance vision. A bear cannot, for example, make out the form of a human at 300 yards. There is ample compensation, however, for mediocre eyesight. Hearing is excellent and the animal can recognise and isolate sounds in the midst of any great hubbub or uproar. Even more highly developed is the sense of smell. The bear continually sniffs the air to gather information.

Because the world of the bear is essentially composed of sounds and scents it might be assumed that the animal makes little use of its mind, but this is not so, as has been shown by scientists who have made intensive tests on the subject. According to one group of experimental studies on comparative animal intelligence the bear is inferior only to several primates, the

Although it is capable of standing upright, the brown bear seldom attacks or defends itself in this position. Its characteristic four-legged method of locomotion is an amble (like that of the camel) and it can attain a fair speed over a short distance.

elephant and dolphins. Dr W. T. Hornaday, using a complex graphic system registering various aspects of behaviour, has claimed that the grizzly bear's intellectual capacity is greater than that of the chimpanzee. In any event it is generally accepted that the bear has an excellent memory, and the ability to plan ahead, its strategy being tempered by extreme prudence.

The secretive lover

The brown bear's mating season commences around mid-June, reaching its peak in July. The strange thing, however, is that whereas the majority of animals exhibit great boldness and passion under the stress of sexual desire, the bear now seems to become more cautious and secretive than at any other time of year so that its courtship ritual is very difficult to observe. Most naturalists affirm that the species is monogamous; at any rate this is true of the Pyrenees population. Furthermore, the male is as a rule faithful to his previous partner. It would appear, however, that if the balance of the sexes is upset by a local preponderance of females, the males will so far modify their normal behaviour as to mate with several partners, ensuring that there is no significant decline in numbers.

The rutting female gives out a distinctive odour and this serves

The newborn bear cub is a tiny, defenceless creature, weighing under 1 lb and measuring only about 9 inches in length. This diminutive size would appear to be an adaptive phenomenon whereby the mother–who is at this time in a semi-lethargic condition and living on reserves of fat–does not become too exhausted either in the process of giving birth or suckling.

as a guide to the reproductive male. Having located her he simply follows her around, continuously emitting low snorts. During the days before mating occurs he licks the muzzle and gently bites the nape, neck and back of his partner, occasionally giving her light cuffs with his paw as signs of his increasing desire. This courtship period is often punctuated by fights with rivals which seldom have serious consequences. Such contests usually take place only between males of approximately equal weights. In other situations thc smallcr animal normally avoids a confrontation by turning tail. Any wounds that are inflicted are of a minor nature and death will only occur accidentally. Olaus J. Murie, however, reports that in Alaska fights between two old grizzlies tend to be exceptionally fierce and that the loser may sustain injuries that will handicap his future activities.

In the Mount McKinley National Park, Alaska, the bears mate between mid-May and mid-June, about a month before it happens elsewhere. The sexual act lasts between from ten to fifteen minutes to several hours. The gestation period is 180–250 days but there appears to be a lapse between impregnation and the start of embryo development. Since the birth takes place during the winter the mother-to-be is far more meticulous than the male in selecting a suitable lair. In the Pyrenees most births occur between 10th January and 20th February and these dates are fairly valid for other regions as well.

The number of cubs in a litter varies according to the age of the mother. Comparatively young females of five to six years will only have one cub, mature females two or three, elderly females one or two. The birth rate of the species is fairly high since each female may have a litter every other year.

The brown bear at home

The newborn cub is so tiny that it is about a month before it even begins to look like a bear at all. It is blind and toothless, with pale, sparse fur. It weighs less than a pound and measures about 9 inches. Since it is so small the mother usually has a quick delivery, losing little blood in the process and at once swallowing the placenta. She tends to her helpless baby's needs with the utmost care and affection, licking it incessantly and holding it in her arms against the thick fur of her chest to keep it warm. The unusually small size of the cubs would seem to be an adaptation to the species' winter habit. If they were larger the suckling process might prove so exhausting to the mother in her weakened condition that the cubs' chances of survival could be jeopardised.

In spring the mother leaves the cave and her cubs to find food. This is the only time that they may be left alone at the mercy of predators such as wolves, foxes, lynxes and wildcats for they are still too small and inexperienced to defend themselves.

The cubs are raised on an exclusive milk diet for about three and a half months although suckling may continue at intervals for a somewhat longer period. But provided the weather is good they will gradually venture out in early spring to experiment with other types of food. This is an important time in their life for

Facing page : The number of cubs in a litter will depend in part on the age of the mother. A first litter will normally consist of only one cub but a fully mature female may give birth to three. Towards the end of her life the number will again be one or two.

they have to learn from the mother how to feed on vegetation and in due course how to hunt.

For the time being the cubs busy themselves unearthing roots, bulbs and tubers, nibbling grass and swallowing slugs, snails and insect larvae. The mother keeps an exceptionally careful watch on their activities. Should they misbehave–perhaps rashly ignoring her warning growl to push past her–she administers a sharp reprimand with a cuff of the paw powerful enough to send the culprit reeling. Only when the cubs are stronger will she permit them to lead the way. Hainard tells how he saw a female brown bear enraged at one of her cubs which had clambered, unnoticed by her, up a tree. Before the youngster had a chance to get down she started shaking the trunk so violently that he tumbled to the ground, there to be cuffed by his mother, though not with any great severity. Intruders also have to be on their guard for the female brown bear protecting her cubs is likely to be aggressive and highly dangerous.

It is interesting to note that although orphans are invariably adopted by a foster mother, adult females have been known to kill their own cubs. This uncharacteristic behaviour has not so far been satisfactorily explained. Are such females mentally disturbed? Is it perhaps a subtle method of controlling the population? Nobody really knows.

By autumn the cubs are sturdy enough to accompany their mother on hunting expeditions and to help her prepare for hibernation. They are fifteen months old when they once more emerge from the winter refuge and the mother can now afford to relax her perpetual vigilance. Towards the end of their second summer they are well on the way to becoming independent. By this time the adult female will be coming into breeding condition again but even after mating, when she is once more gravid, she will maintain periodic contact with her cubs, who will spend the ensuing winter in a lair very close to hers.

By the following spring the bears weigh between 60 and 90 lb and males and females separate, though remaining in groups for a while longer. At three and a half years they are adults and able to procreate. They still spend much time in playful activity, such games being marked by considerable ingenuity and imagination. The young bears may, for example, divert themselves by throwing and catching pieces of wood or pine cones. Some have even been seen converting a tree trunk which has fallen across a rock into a makeshift see-saw.

The adults also show playful instincts at times and the fact that they do not resort to stereotyped movements, as do other mammals, but tend to improvise freely according to circumstances furnishes additional proof of their intelligence. Thus, bears in mountain regions have sometimes been observed rolling down steep slopes, head tucked between paws, only to clamber up the hill once more, take a brief rest at the summit and repeat the process again and again. At other times groups of bears have been seen ambling along in customary single file when suddenly one of them, probably the last in line, will start making repeated efforts to trip up the animal in front. The latter may stumble or even fall, without growling or showing any other sign of ill humour.

Some zoologists claim that a cub of under two years old, described as a 'mentor', will sometimes 'hibernate' with its mother and help her to rear the new litter, but this phenomenon still lacks positive confirmation.

The adult male certainly plays no part in caring for his cubs and social life as such is virtually non-existent. Groups of adults are sometimes seen together in a particular locality, with one male behaving in a dominant manner and perhaps expelling others from his temporary domain. But territory is not normally an important consideration and although fights do sometimes occur, mutual tolerance is the rule. In this connection it is amusing to note how adult bears in North America treat rivals when engaged in fishing for salmon. The fishes fight their way up-river to the spawning grounds in such enormous numbers that there is no point in a bear aiming to prevent a companion trying its luck in the same turbulent waters. He will therefore graciously tolerate the presence of a rival by pretending to take no notice of him. Within a comparatively small area, therefore, several bears may be observed studiously devoting themselves to the matter in hand without paying the slightest attention to one another. If two animals do happen to come nose to nose they simply turn aside disdainfully and continue fishing.

Young brown bears spend much of their time playing games, these activities being characterised by much improvisation and originality–signs of advanced intelligence.

Facing page: The female brown bear patiently attends to the needs of her cubs and supervises their activities until they are big enough to find their own food and defend themselves efficiently.

FAMILY: Ursidae

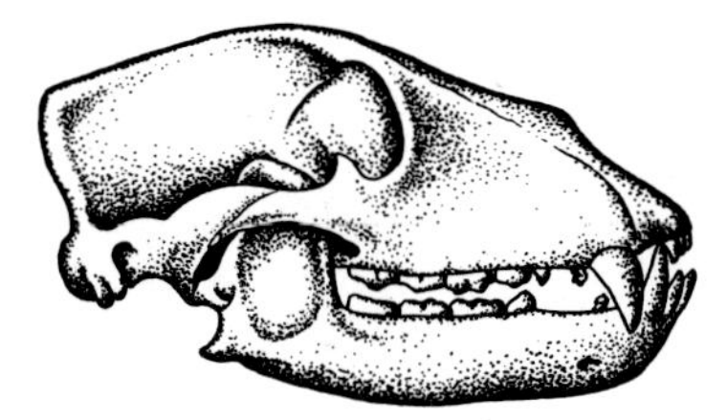

Skull of the brown bear.

Although it is a simple matter to distinguish a bear from any other carnivorous or omnivorous mammal, zoologists have encountered difficulties in trying to separate the bear family (Ursidae) scientifically from the dog family (Canidae) as a result of resemblances between prehistoric fossils of either group. In an attempt to resolve this point George Gaylord Simpson, in 1931, linked the two in a single superfamily (Canoidea).

The Ursidae are large carnivores with a massive body and a rudimentary tail. They are plantigrades and their toes are furnished with non-retractile claws.

Theoretically the dentition is:

$$I: \frac{3}{3} \quad C: \frac{1}{1} \quad PM: \frac{4}{4} \quad M: \frac{2}{3}$$

This gives a total of 42 teeth but in fact the number of premolars is frequently less. Thus the extant brown bears possess 36–38 teeth, so that their dentition formula is as follows:

$$I: \frac{3}{3} \quad C: \frac{1}{1} \quad PM: \frac{3}{2\text{–}3} \quad M: \frac{2}{3}$$

The various groups of bears are so similar in appearance and behaviour that it is virtually impossible to classify them under separate subfamilies, so slender are the differences between them. In some cases it is just as difficult to distinguish individual genera and species. Thus the American black bear (*Euarctos americanus*) resembles the brown bear (*Ursus arctos*) so closely that many authors prefer to regard both animals as members of the genus *Ursus*.

With the exception of the polar bear members of the Ursidae are typically inhabitants of temperate or tropical regions. All of them–but especially the brown bear and polar bear–have been extensively hunted. Nevertheless their numbers are still considerable. Bears are found throughout the northern hemisphere with two species ranging south of the equator.

Latest classification divides the Ursidae into seven genera, the most numerous and diversified of which is the genus *Ursus*. The most familiar species, often represented in zoos, is the Eurasian brown bear and some authors consider that other members of the genus are simply subspecies of *Ursus arctos*. The Syrian brown bear (*Ursus arctos syriacus*) is generally accepted as a subspecies, as are the red or isabelline bear (*U. a. isabellinus*) and the blue or snow bear (*U. a. pruinosus*), all of them inhabitants of Asia. The North American grizzly bear (*Ursus horribilis*) is likewise found in various forms and the Kodiak bear (*Ursus middendorfii*) is considered by some authorities to be a subspecies of the grizzly.

The American black bear is the only species belonging to genus *Euarctos* and the main distinguishing feature between this animal and the brown bear is the shape of its muzzle. Like the latter it is normally shy and unaggressive and is protected in a number of national parks.

The genus *Selenarctos* also consists of a single species–the Asiatic or Himalayan black bear (*Selenarctos thibetanus*). Its range of distribution, bordered to the north by that of the brown bear, extends from Iran and through the Himalayan mountain chain as far east as China and Japan. The distinguishing feature of this bear is a V-shaped white patch on the chest, a white upper lip and short black claws. Its diet is more markedly vegetarian than that of other members of the family but like its relatives it also feeds on large numbers of insects and other small or medium-sized invertebrates as well as on honey.

The typical representative of genus *Tremarctos* is the spectacled bear

Facing page: The American black bear closely resembles the brown bear. Its muzzle, however, is longer, bulging slightly on top. The feet too are smaller and the claws less powerful.

Brown bear
(*Ursus arctos*)

American black bear
(*Euarctos americanus*)

Sun bear
(*Helarctos malayanus*)

Himalayan black bear
(*Selenarctos thibetanus*)

Spectacled bear
(*Tremarctos ornatus*)

Honey bear
(*Melursus ursinus*)

The representatives of six of the seven genera of Ursidae all show physical resemblances and inhabit temperate and tropical regions mainly. The polar bear, not depicted here, belongs to the genus *Thalarctos* and is distinguished by its white colour and Arctic habitat.

(*Tremarctos ornatus*), a comparatively rare species living in the Andes. It is a small, agile animal with black or brownish-black fur. The common name is derived from the white patches surrounding the eyes.

The fifth genus, *Helarctos*, is represented by the Malayan or sun bear (*Helarctos malayanus*) from the forests of South-east Asia. The smallest of all bears, with black hair, yellowish muzzle and a similarly coloured patch on the chest, it is a nocturnal, tree-dwelling species. There is, however, considerable variation in the white markings on the face.

The sole member of genus *Melursus* is the sloth or honey bear (*Melursus ursinus*) from the jungles of India and Ceylon. Its fur is shaggy and black with a white patch on the chest. The teeth are poorly developed and well suited to a basically insectivorous diet. The lips are particularly long and mobile, useful for sucking up termites.

The seventh genus, *Thalarctos*, differs radically from the others for it comprises the largest and most carnivorous of all Ursidae, the polar bear (*Thalarctos maritimus*). The fur is white all over, often tinged with yellow, providing perfect camouflage in the animal's Arctic surroundings. Seals are the polar bear's favourite prey.